Topics in
Biological Inorganic Chemistry

Volume 1

Springer

Berlin
Heidelberg
New York
Barcelona
Hong Kong
London
Milan
Paris
Singapore
Tokyo

Metallopharmaceuticals I

DNA Interactions

Editors: M.J. Clarke · P.J. Sadler

With contributions by
E. Alessio, C.F.J. Barnard, A. Bergamo, K. Borsky,
M. Coluccia, N.P. Farrell, A. Gelasco, L.R. Kelland,
B. K. Keppler, M. Leng, S.J. Lippard, B. Lippert,
G. Mestroni, G. Natile, T. Pieper, F.I. Raynaud,
J.D. Roberts, G. Sava

 Springer

Volume Editors:

Professor Michael J. Clarke
Department of Chemistry
Boston College
Merkert Center
Chestnut Hill, MA 02167
USA

Professor Peter J. Sadler
Department of Chemistry
University of Edinburgh
King's Building
West Mains Road
Edinburgh EH 3JJ
Scotland, GB

ISSN 1437-7993
ISBN 3-540-64889-5 Springer-Verlag Berlin Heidelberg New York

Library of Congress Cataloging–in–Publication Data applied for
Die Deutsche Bibliothek - CIP-Einheitsaufnahme
Metallopharmaceuticals : DNA interactions / with contributions by E. Alessio ... -
Berlin ; Heidelberg ; New York ; Barcelona ; Hong Kong ; London ; Milan ; Paris ;
Singapore ; Tokyo : Springer, 1999
  (Topics in biological inorganic chemistry ; Vol. 1)
  ISBN 3-540-64889-5

© Springer-Verlag Berlin Heidelberg 1999
Printed in Germany

Cover: Friedhelm Steinen-Broo, Pau/Spain; MEDIO, Berlin
Typesetting: Data conversion by MEDIO, Berlin

SPIN: 10538568   31/3020 - 5 4 3 2 1 0 - Printed on acid-free paper.

Editorial Board of the Series

Preface to the Series

Biological Inorganic Chemistry is at the dynamic interface of Chemistry and Biology. Both research and teaching in biological inorganic chemistry are increasing at an enormous rate, aided by advances in biophysical, synthetic, biochemical and molecular biological concepts and technology. Publications describing original research, reviews articles, symposia and textbooks are appearing with increasing frequency in the traditional venues for inorganic chemistry, biochemistry and molecular biology. Enormously successful, small specialists' meetings have been happening for decades.

The critical mass for a Society representing the field of Biological Inorganic Chemistry was achieved a few years ago at the seventh meeting of the *International Conference on Biological Inorganic Chemistry* (ICBIC) in 1995 when The Society of Biological Inorganic Chemistry (SBIC) was founded. ICBICs had already increased in size over ten-fold from the first meeting in 1983. As the millenium approaches, SBIC is a healthy young society which owns a strong new publication, the *Journal of Biological Inorganic Chemistry* (JBIC). The Chief Editor of JBIC is Ivano Bertini. With JBIC Volume 5, published in the year 2000, Lawrence Que will become JBIC Chief Editor and Ivano Bertini, Founding Chief Editor.

JBIC's success, attested to by the quality of the articles, the journal publication rate, institutional subscriptions, impact factors and by students who say there is finally a journal for "Our Field!", has prompted the JBIC publisher, Springer, to launch a series of thematic reviews in TBIC, *Topics in Biological Inorganic Chemistry*. The first two TBIC inaugral volumes will be dedicated to Metallopharmaceuticals and are edited by Michael J. Clarke and Peter Sadler. As SBIC president during 1998–2000, I speak for all the members and lead with the benefit of the wisdom and foresight of all the early SBIC officers and supporters, in particular David Garner, Ivano Bertini, Harry Gray, Kenneth Raymond, Joann Sanders-Loehr, Robert Scott, Roger Thorneley, Karl Wieghardt and Antonio Xavier. SBIC welcomes TBIC as the new Springer complement to JBIC and to all the BIC activities.

Oakland, CA, March, 1999   Elizabeth C. Theil

Preface to Volume 1

The resounding success of the simple transition metal complex cisplatin against some types of cancer has led to a number of important developments in the use of metal ions as chemotherapeutic agents. Nevertheless, in the two decades since the introduction of cisplatin into the clinic, no chemotherapeutic drugs involving other transition metals have achieved widespread use. Gelasco and Lippard begin the volume with a survey of the history and development of cisplatin and related agents. Combining insights from both coordination chemistry and molecular biology, they review likely molecular biological mechanisms initiated by the binding of cisplatin to DNA. Their own work suggests that a cisplatin-induced bend in DNA results in an increased affinity for a class of DNA binding proteins that may initiate a suicidal response in the cell. Recognition of the platinum lesion by DNA repair enzymes also allow cancer cells to develop resistance to the drug. Barnard, Raynaud and Kelland provide an interesting account of the discovery of an orally active anticancer drug containing Pt(IV), which is now in phase II clinical trials and may offer promise against cisplatin-resistant cancers.

While the cis-geometry of cisplatin was once thought to be a structural requirement for anticancer activity, Natile and Coluccia describe the efficacy of a number of Pt(II) and Pt(IV) complexes with trans geometries, which probably open up new modes of crosslinking DNA.

Farrell, Qu and Roberts present a new class of Pt-antitumor agents that also extends the possibilities of cross linking DNA. In this case, two or more Pt(II) centers are tethered together, thereby lengthening the reach of the crosslink. The first of these polynuclear drugs is now in phase I clinical trials. As a class, these are structurally diverse agents, in which modifications might be made to target different types of cancers.

Lippert and Leng provide an overview of platinum-oligonucleotides as a new type of blocking agent in either antisense or antigene approaches. These could be regarded as analogs of polynuclear agents, but with the tethered oligonucleotide providing enormous selectivity for directing the Pt to a particular nucleic acid sequence.

The broad range of possibilities for other elements to mount effective attacks on cancers is perceptively reviewed by Pieper, Borsky and Keppler. Their overview highlights how other metal ions can hijack transferrin, the protein vehicle for transporting iron in blood, to gain entry into tumor cells. Importantly, the panorama of mechanistic possibilities they present could suggest new approaches to cancer chemotherapy. Finally, Sava, Alessio, Bergamo and Mestroni outline their studies on ruthenium dimethyl sulfoxide complexes that bind to DNA. Surprisingly, the most promising of these, which is about to enter clinical trials, is not active by binding to DNA nor does it attack the primary tumor. Rather, it generates an antimetastatic effect by inhibiting a proteinase resulting in a thicker matrix surrounding the tumor and its blood vessels.

Overall, this volume should provide bioinorganic chemists, molecular biologists and oncologists with fresh insights for developing new approaches for using transition metals in chemotherapeutic agents.

March 1999

Michael J. Clarke
Peter J. Sadler

Contents

Anticancer Activity of Cisplatin and Related Complexes

Andrew Gelasco[1], Stephen J. Lippard[2]

Department of Chemistry, Massachusetts Institute of Technology, Cambridge, MA 02139, USA
E-mail: [1] gelasco@lippard.mit.edu, [2] lippard@lippard.mit.edu

Cisplatin is a simple inorganic compound which has proved to be a very effective anticancer drug, especially against testicular tumors. The compound has therapeutic limitations based primarily on toxic side effects and acquired resistance. The mechanism of action of cisplatin is under intense study in order to understand the chemical basis for these limitations and to develop in a rational manner a new generation of platinum-based antitumor drugs. These studies have led to the discovery of new classes of anticancer platinum compounds, including some which utilize both mononuclear and dinuclear Pt(II) and *trans*-Pt(IV) geometries. High-resolution structures of cisplatin and derivatives bound to duplex DNA have been determined both by X-ray crystallography and by NMR solution methods. Also noteworthy is the discovery of DNA-binding proteins that specifically recognize cisplatin-modified DNA. The contributions of each of these different areas of study towards an understanding of the molecular mechanism of the anticancer activity of cisplatin are discussed.

Keywords. Cisplatin, Anticancer, DNA, HMG-domain proteins, Platinum drugs

List of Abbreviations

DDP   diamminedichloroplatinum(II)
HMG   high mobility group
NMR   nuclear magnetic resonance

NOESY   nuclear Overhauser effect spectroscopy
PECOSY   purged exclusive correlated spectroscopy
RMSD   root mean-squared deviation
TEMPO   2,2,6,6-tetramethylpiperidinyloxy
rMD   restrained molecular dynamics
en   ethylenediamine
PAGE   polyacrylamide gel electrophoresis
SDS   sodium dodecyl sulfate
NER   nucleotide excision repair

1
Introduction

Cisplatin [*cis*-diamminedichloroplatinum(II) or *cis*-DDP] is one of the most widely used and successful drugs in cancer chemotherapy [1]. Highly effective against a number of tumors [2], its role in curing testicular cancer is well established [3]. In combination therapy with other drugs (e.g. bleomycin, doxorubicin or 5-fluorouracil), cisplatin is used to treat tumors of the head and neck, ovaries, cervix, and bladder. Treatment of these cancers is often limited by the recurrent formation of cisplatin-resistant tumors [4]. Because of this and other limitations, understanding the mechanism and reducing the toxicity of cisplatin are important for developing more effective drugs based on platinum chemotherapy.

1.1
History of Cisplatin as an Anticancer Drug

The biological properties of cisplatin were discovered serendipitously during experiments studying the effect of electric fields on the growth of *E. coli* bacteria [5]. Platinum electrodes used in the experiment reacted with the bacterial growth media to produce Pt(II) and Pt(IV) compounds, including *cis*-[Pt(NH$_3$)$_2$Cl$_2$], which inhibited cell division causing the bacteria to become long and filamentous. Since inhibition of cell division has often been associated with anticancer agents, the platinum compounds were examined for antitumor activity in mice [6]. Based on these studies *cis*-diamminedichloroplatinum(II), now known as cisplatin, was found to be very effective against a variety of solid tumors, whereas its *trans*isomer (*trans*-DDP) was inactive. The structures of cisplatin and *trans*-DDP are shown in Fig. 1.

cisplatin   *trans*-DDP   **Fig. 1.** Cisplatin and its *trans*isomer, *trans*-
(*cis*-DDP)   DDP

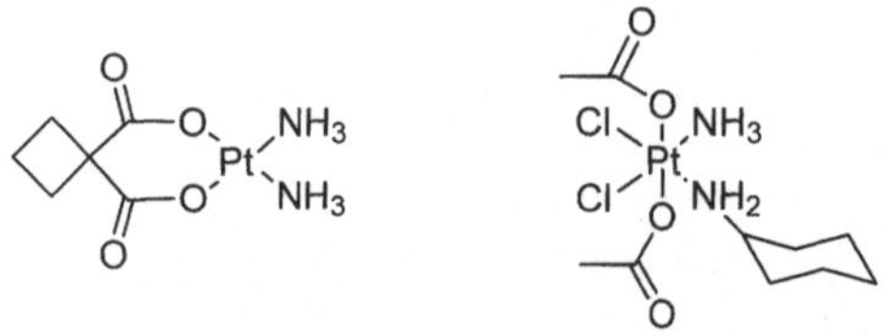

carboplatin   JM216

Fig. 2. Second-generation cisplatin derivatives carboplatin and an oral analog

When cisplatin initially entered clinical trials, the high levels of nephrotoxicity observed during drug administration almost prevented its further development. Prehydration therapy with intravenous saline, however, and controlled dosage levels of less than 120 mg/m^2, greatly reduced renal toxicity as a side effect of cisplatin treatment. Continued studies led to FDA approval of the drug in 1978 as a treatment for genitourinary tumors.

Even while cisplatin was entering the clinic for treatment against primarily testicular cancer, studies were ongoing to combat the numerous side effects of its chemotherapy, which included nausea, vomiting, neurotoxicity, and nephrotoxicity. These early studies resulted in a second-generation platinum drug, diammine(cyclobutanedicarboxylato)platinum(II), commonly referred to as carboplatin (Fig. 2). This compound was presumed to have the same active form of the drug as cisplatin, but the slower hydrolysis of the cyclobutanedicarboxylate leaving group of carboplatin allows a similar antitumor activity with reduced toxicity [7]. Carboplatin is now routinely used as an alternative to cisplatin in chemotherapy.

Whereas both cisplatin and carboplatin are administered intravenously, a new class of Pt(IV) compounds has been prepared that can be taken orally. These compounds, one of which is currently undergoing clinical trials, have the general structure cis-[Pt(amine)$_2$Cl$_2$(RCH$_2$COO)$_2$]$_2$. An example is shown in Fig. 2 as JM216 [8]. The mechanism of action of these octahedral complexes is still under study, but most likely they are reduced to a bis(amine)dichloroplatinum(II) species which would then react like cisplatin in vivo [9]. These compounds are discussed in more detail in Sect. 2.1.2.

When cisplatin is administered by intravenous injection, the neutral form of the drug can easily diffuse into individual cells. The relatively high (~150 mM) serum chloride ion level inhibits conversion to the hydrolyzed cis-[Pt(NH$_3$)$_2$(H$_2$O)$_2$]$^{2+}$ form. The much lower chloride ion concentration inside the cell leads to the formation of first the mono- and then diaquadiammineplatinum(II) complexes. These activated species can then react with a variety of intracellular macromolecules including DNA, RNA, and proteins. The reaction of cis-[Pt(NH$_3$)$_2$(H$_2$O)$_2$]$^{2+}$ with DNA affords a variety of adducts; the major ones are depicted in Fig. 3.

In vitro studies of platinated oligonucleotides by enzymatic digestion analysis has indicated that the major product formed is the 1,2-intrastrand cross-link

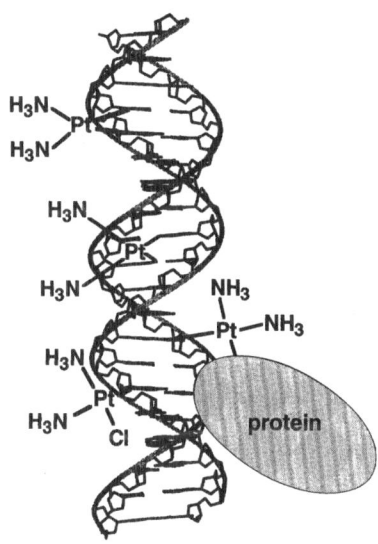

Fig. 3. Schematic representation of the distribution of cisplatin adducts on duplex DNA, the 1,2-*intra*strand, GG-*inter*strand, monofunctional, and protein-DNA cross-links. The chloride depicted could also be a water molecule.

between the N7 positions of adjacent purine bases. These adducts [Pt(NH$_3$)$_2$ {d(GpG-N7(1),N7(2))}] and [Pt(NH$_3$)$_2${d(ApG-N7(1),N7(2))}], account for ~65 and ~20% of platinum on DNA, respectively [10]. Other minor products, including the *inter*strand G-Pt-G adduct formed at GC sites and monofunctional Pt-G adducts, together account for less than 5% of the total. A study of human cells in culture has revealed a very similar spectrum of Pt-DNA adducts formed in vivo, as detected by immunochemical techniques [11].

1.2
Early Structural Studies of Cisplatin-Nucleotide Complexes

Understanding the mechanistic basis for the anticancer activity of cisplatin requires knowledge of the structural consequences of the Pt-DNA adducts formed upon drug binding. Early structural information about the major cisplatin adduct, the bifunctional *cis*-[Pt(NH$_3$)$_2${d(pGpG)}] *intra*strand cross-link, was obtained by X-ray crystallography [12], validating preliminary NMR solution studies on the related complex [Pt(NH$_3$)$_2${d(GpG)}]$^+$ [13]. This early crystal structure revealed details of platinum coordination to the guanosine bases through the N7 atoms, resulting in a head-to-head base conformation. As shown in Fig. 4, the {Pt(NH$_3$)$_2$}$^{2+}$ moiety causes a significant disruption of the stacking between the adjacent guanine bases, with a resulting dihedral angle between the base planes of 87°. In addition, the 5'-ribose ring switches from the S to the N conformation and one of the Pt-ammine ligands forms a hydrogen bond with the 5'-terminal phosphate oxygen atom. The formation of this hydrogen bond may explain the requirement for amine ligands on active platinum antitumor drugs. A

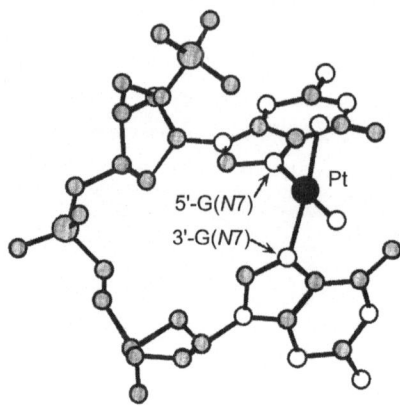

Fig. 4. Representation of the X-ray crystal structure of cisplatin bound to the dinucleotide d(pGpG), its major adduct in DNA, redrawn from coordinates in [45]

very similar conformation occurs in the cisplatin adduct of the trinucleotide CpGpG, which has the cis-[Pt(NH₃)₂{d(GpG-N7(1),N7(2))}] chelate and for which the 5'-cytidine base is involved in intermolecular interactions in the solid state [14].

A series of NMR spectroscopic studies of 1,2-intrastrand cross-links in platinated duplex DNA has been carried out [15,16]. Structures determined by using a combination of molecular mechanics and dynamics calculations afforded models consistent with the observed NMR data. These models indicate that the 1,2-intrastrand d(GpG) chelate formed in the duplex oligomer results in a kink towards the major groove, but that otherwise the DNA remains undistorted. Gel electrophoresis studies of cisplatin-modified duplex DNAs subsequently showed that the helix was bent by ~40° towards the major groove [17,18]. Furthermore, the DNA helix was also distorted by unwinding near the site of the platinum adduct [19], indicating that the effect of cisplatin binding to DNA was rather complex and that various structural factors may be responsible for the anticancer activity of the drug. These early observations stimulated considerable further investigations into the detailed nature of the structural perturbations caused by the major cisplatin adduct on duplex DNA and are discussed later in this review.

1.3
Scope of This Review

Although carboplatin and JM216 have proved to be effective in modulating the side effects and inconvenience of cisplatin chemotherapy, the development of new platinum-based drugs continues in order to improve the clinical activity of this class of antitumor compounds. These studies have been complemented by high-resolution structural studies of cisplatin-duplex DNA adducts and investigations into protein-cisplatin-DNA interactions. The present article surveys these developments over the last five years and examines future prospects for the development of platinum-based anticancer drugs. Section 2 focuses on recent advances in platinum drug design and efforts to circumvent resistance to cisplatin and carboplatin. Studies of the mechanism of cisplatin binding to both single-stranded and duplex DNA are summarized in Sect. 3. Section 4 explores a number of recent high-resolution structure determinations of platinum adducts on duplex DNA and describes the distortions of the helical structure of DNA caused by the different platinum compounds. Finally, Sect. 5 discusses the role that a class of cellular proteins, the HMG-domain proteins, may have in the cytotoxicity of cisplatin and related compounds.

2
Development of Cisplatin Derivatives

2.1
Platinum(II) and Platinum(IV) Amine Derivatives

Although cisplatin and its closely related analog carboplatin are two of the most effective and widely used antitumor drugs, there are limitations to their applicability due to the high level of resistance observed in various tumors [20]. There have been three major mechanisms proposed for cisplatin resistance in vivo [21]. First, the intracellular concentrations of the drug may be decreased either by blocking uptake or increasing efflux. Second, the drug can be inactivated by quenching the monofunctional adducts before they form cross-links, through reaction with other ligands in the cell, or by reaction with metallothioneins, glutathione or other intracellular ligands. Finally, removal of drug-DNA adducts may occur by cell-based repair systems including nucleotide excision repair. The need to circumvent these processes has driven a search for analogs which block the cross-resistance commonly observed for cisplatin but achieve a similar level of effectiveness. One of the problems associated with the development of such derivatives has been the production of compounds having higher nephrotoxicity or unanticipated mutagenicity.

Derivatives in which one of the amine ligands has been modified have produced some new platinum(II) compounds with clinical potential and platinum(IV) compounds which may allow the use of orally administered platinum drugs. Figure 5 shows the structures of several cisplatin derivatives that have been designed recently, details of which are discussed below.

2.1.1
Platinum(II) Amine Derivatives

One strategy to overcome cisplatin (and carboplatin) resistance has been to use sterically hindered carbocyclic amines such as the pyridine-amine derivatives shown in Fig. 5, including DWA2114R [22], AMD508, AMD494 and most recently AMD473 [23,24]. Compounds such as DWA2114R are believed to form DNA adducts similar to those made by cisplatin, but the bulky ligand may inhibit nu-

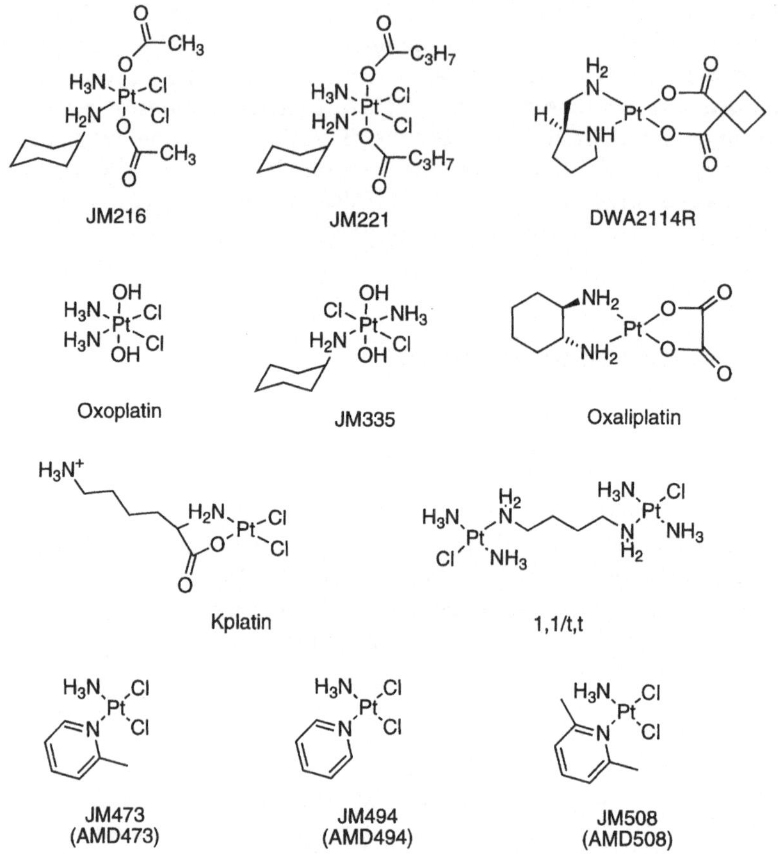

Fig. 5. Third-generation cisplatin derivatives which were investigated for their antitumor activity

cleotide excision repair [22]. The use of the chiral ligand in DWA2114R provides the interesting observation that the *R*-isomer is not toxic while the *S*-isomer (shown in Fig. 5) is [22]. The activity of a chiral platinum drug may provide insight into the requirements of adduct formation or possibly the intermolecular interactions that occur following DNA binding (vide infra). This compound is currently in phase III clinical trials [25].

The second class of sterically hindered compounds are depicted across the bottom of Fig. 5. The pyridine-derived ligands retard hydrolysis of the chloride ions relative to cisplatin. It has also been suggested that the bulkier ligand may inhibit the reactivity of the platinum drug with glutathione and other sulfur donors in vivo, thus circumventing a proposed mechanism of acquired resistance [23]. In vitro studies of AMD473 have shown diminished cross-resistance to cisplatin in three different cell lines with acquired cisplatin resistance [24]. The binding of AMD473 to salmon sperm DNA was much less affected than cisplatin binding in the presence of 5 mM glutathione, indicating that this drug may indeed avoid this type of acquired resistance. This compound is reported to have entered phase I/II trials in England in 1997.

Oxaliplatin [(1,2-*trans*-diaminocyclohexane)oxalatoplatinum(II), L-OHP] has been undergoing clinical trials since 1986, both in combination therapy and as a single agent drug for the treatment of colorectal cancer [26,27]. Again the use of a bulky platinum ligand, diaminocyclohexane, is postulated to prevent recognition and repair of platinum adducts on DNA similar to those formed by cisplatin. This compound, which contains the oxalato leaving group, is somewhat effective in patients who have platinum-resistant ovarian cancer [27], and was the first cisplatin derivative to be approved in France for the treatment of advanced colorectal cancer [26].

2.1.2
Platinum(IV) Amine Derivatives

Another effort to combat the intrinsic resistance to platinum drugs has focused on the development of Pt(IV) bis(amine) compounds. Two cisplatin analogs, JM216 [8] and JM221 [28–30] (see Fig. 5), have been investigated. Preclinical evaluation of JM216, *cis,trans,cis*-[PtCl$_2$(OAc)$_2$NH$_3$(*c*-C$_6$H$_{11}$NH$_2$)], in ovarian cell lines indicated similar cytotoxicity to cisplatin (IC$_{50}$ of 1.7 μM for JM216 vs. 3.5 μM for cisplatin) [8]. In ovarian, testicular and cervical cell lines representing classes of both acquired and intrinsic cisplatin resistance, JM216 exhibited no cross-resistance. This compound is especially effective in cell lines where resistance is based on decreased drug accumulation [8]. It is currently in clinical trials [31] and has been used as a lead compound towards the development of JM221, a third-generation platinum(IV) drug with a very similar geometry to JM216.

JM216 is active when administered orally [32]. This method of drug delivery is appealing since it would free the patient from intravenous administration and in turn lower the cost of treatment. The effectiveness of these compounds in both resistant and sensitive cell ovarian cell lines again showed no cross-resistance to cisplatin. In a study of cisplatin and JM221 with a cell line exhibiting intrinsic cisplatin resistance, the mechanism of both the resistance to cisplatin and effectiveness of JM221 was examined [28]. The in vitro model showed that the cisplatin resistance was due primarily to inhibition of bifunctional adduct formation, but that JM221 circumvented the observed resistance through improved drug uptake and by decreased repair.

For both JM216 and JM221, the narrow spectrum of adduct formation has suggested that the drug is activated by reduction to the Pt(II) form, which gives a similar distribution of products to that observed for cisplatin as revealed by in vitro platination studies [9,28]. The Pt(IV) compound oxoplatin (Fig. 5) binds to DNA targets without the addition of external reducing agents [30]. Antibodies to *cis*-diammineplatinum(II)-DNA lesions were not sensitive to the adducts formed by oxoplatin. After addition of an exogenous reducing agent, however, the platinated DNA was sensitized to the antibodies. These results imply that a Pt(IV) adduct forms initially which differs in shape from those formed by cisplatin but which can be converted to analogous Pt(II) adducts. Although the reactions of platinum(IV) complexes are inherently slower than those of platinum(II), the discovery of the direct formation of Pt(IV) octahedral DNA adducts would open up this area of study for such compounds.

2.2
Non-Amine-Based Platinum(II) Derivatives

Recently, combinatorial methods were applied in an effort to produce platinum(II) amino acid complexes (Pt-AA) having potential antitumor activity [33]. The complexes were synthesized by allowing various mixtures of amino acids to react with $K_2[PtCl_4]$ in water. Initially, a mixture of 17 amino acids (excluding H, C, and M) was allowed to react at a 2:1 molar ratio with $K_2[PtCl_4]$. The resulting mixture of compounds was then used to modify a 123 bp fragment of DNA and the resulting adducts screened for their ability to bind to HMG1 in a gel mobility shift assay. HMG1 is the prototype for a class of DNA binding proteins known as HMG-domain proteins, which bind selectively to cisplatin-modified DNA [34,35]. They recognize the structural deformations induced upon formation of 1,2-*intra*strand cross-links at adjacent purine sites on duplex DNA [36]. Because HMG-domain proteins can modulate the cytotoxicity of cisplatin (vide infra), HMG1 was used to screen mixtures of Pt-AA complexes in order to select those having potential cytotoxic properties [33].

Randomly prepared sub-libraries containing four amino acids each were prepared, and gel mobility shift assays were used to isolate Pt-AA complexes which enhanced DNA-protein binding. Successive library selections identified the lysine complex $[PtCl_2(lysine)]$ as the most active compound [33]. The complex, referred to as Kplatin, was prepared by adding one molar equivalent of lysine to $K_2[PtCl_4]$. An X-ray crystallographic study of Kplatin revealed the novel N,O-chelated platinum structure shown in Fig. 5. Studies of the cytotoxicity of Kplatin indicated only marginal toxicity towards human tumor cell lines, however, with a typical LC_{50} value of 60 µM in HeLa cells compared to 0.5 µM for cisplatin. Nevertheless, this compound represents a lead in a combinatorial methodological approach to the discovery of new platinum drugs, which has recently been reviewed in more detail [37].

2.3
trans-Platinum(IV) Complexes

The *trans,trans,trans*-ammine(cyclohexylamine)dichlorodihydroxoplatinum(IV) complex JM335 has comparable cytotoxicity to cisplatin in vitro against five different ovarian carcinoma cell lines (average IC_{50} of 3.1 µM vs. 4.1 µM for cisplatin) [38] and displays non-cross-resistance in five of seven cisplatin-resistant cell lines [38]. In studies of the cisplatin-resistant cell line, the *cis,cis,trans*-isomer of

JM335, known as JM149, showed a similar cross-resistance pattern to that observed for cisplatin but was very different from JM335. In contrast to *trans*-DDP, which is believed to have low cytotoxicity due to its inability to form the major cisplatin 1,2-*intra*strand adduct, JM335 forms such adducts as well as the *inter*strand cross-link, also formed by both cisplatin and *trans*-DDP. The reasons for the observed cytotoxicity of JM335 are still under investigation, but like the related compounds JM216 and JM221, which are in clinical trials, the strength of this compound may be its effectiveness in overcoming resistance in cells where resistance derives from decreased drug uptake.

Additional investigations of a series of *trans*-platinum complexes based on JM335 has indicated that this class of complexes has both in vitro and in vivo antitumor activity [39]. These derivatives have the general formula *trans,trans, trans*-[PtCl$_2$NH$_3$(NH$_2$R)(OH)$_2$]. In most cases (12 of 14), the *trans*analogs showed equal or better cytotoxicity than their *cis*-isomers. This large number of similar derivatives has provided additional insight into how these *trans*complexes circumvent cisplatin cross-resistance. In this study the *trans*complexes overcame acquired cisplatin resistance in cell lines where the mechanism of resistance was either due to diminished drug uptake or to enhanced levels of repair [39]. Whereas the bulky ligand on the Pt(II) complexes allowed circumvention of drug inactivation based cisplatin resistance (glutathione interactions, etc.) as described in Sect. 2.1.1, this *trans*-Pt(IV) class has been successfully targeted at the other two major forms of acquired resistance.

2.4
Dinuclear Platinum(II) Derivatives

If the mode of resistance to cisplatin is efficient repair of specific adducts, then an effective means of circumventing such resistance would be to employ platinum complexes that form very different types of DNA lesions. One such approach has been to prepare dinuclear platinum complexes having the general formula [{PtCl$_x$(NH$_3$)$_{3-x}$}$_2$(diamine)]$^{2(2-x)+}$, where x=0–3 and the diamine comprises 1–4 methylene groups (n) per chain. These complexes can potentially form either mono-, di- or trifunctional adducts depending on the value of x. An example of one such dinuclear platinum complex, where x=1 and n=4, is shown in Fig. 5 and labeled 1,1/t,t. Early studies of these compounds employed the difunctional (x=2) series [{*cis*-PtCl$_2$(NH$_3$)}$_2$(NH$_2$(CH$_2$)$_n$NH$_2$)], where n=4–9 [40,41]. These compounds showed good activity against cisplatin-resistant murine solid tumor cell lines in vitro, and the activity appeared to depend on the backbone chain length (n) [40]. The elevated activity level in cisplatin-resistant cells, and in some cases the different activity spectrum, suggests that these compounds have a potentially different mechanism than that proposed for mononuclear cisplatin-type derivatives. Circular dichroism studies indicate that these complexes formed lesions on the DNA similar to those of cisplatin [41]. More recently, however, the focus has been on complexes containing monofunctional platinum atoms, where x=1 [42–44].

These dinuclear compounds were examined for their DNA binding properties by using an enzymatic assay in which the Pt-DNA adduct was first isolated by use of a 3'–5' exonuclease. Resulting DNA fragments containing an *inter*strand cross-link were then elongated using DNA polymerase [42]. This assay revealed that the dinuclear complex [{*trans*-PtCl(NH$_3$)$_2$}$_2$(NH$_2$(CH$_2$)$_4$NH$_2$)]Cl$_2$ formed 1,2-, 1,3- and 1,4-*inter*strand cross-links between guanosine residues on opposite DNA strands. The study also showed that cisplatin formed the expected 1,2-*intra*strand cross-link but that, at certain sequences, in particular GCGG, the *inter*strand and some 1,3-*intra*strand adducts were formed in preference to the 1,2-*intra*strand adduct. Taken together, these results indicate that the dinuclear compounds have both a very different binding mode and also increase the number and type of adducts formed per platinum compound. The longer target sequences of the dinuclear complexes, a 3 base pair span in the case of the 1,4-*inter*strand adduct, has led to continued development of these complexes and suggests reasons for the different repair response that these *inter*strand adducts may have [40].

These dinuclear complexes also form 1,2-*intra*strand adducts [43] on oligomers with discrete –TGGT– sites. Gel retardation assays suggest a flexible non-directional bend in the DNA [43], very different from that observed for the cisplatin adduct, which has a distinct bend toward the major groove [17,18] (see Sect. 4.1). Whereas 2D NMR studies and chemical reactivity studies indicate that this adduct causes local distortion of the bases at the site of platination, it also increases the flexibility of the duplex over a region of five or six base pairs [43]. Such flexibility may be due to the presence of a linker chain between platinum atoms relieving the strain on the adjacent guanosine residues, although it has been shown to be independent of the length of the chain. The NMR structural work also indicated that the local distortions caused by the dinuclear complex adducts were different from those observed for cisplatin on either single-stranded [14,45] or double-stranded DNA [46,47]. The sugar puckers and the base orientations for the –TG*G*T– sequence were N, S/N, N, S, and *anti,anti,anti/syn,anti* in the dinuclear complex vs. S/N,N,S,S and *anti,anti,anti,anti* for cisplatin adducts [46–49]. These subtle differences in local geometry may contribute to the different overall duplex distortion that was observed by gel electrophoresis.

Whereas many of the different cisplatin derivatives discussed showed reasonable activity against cisplatin-resistant cell lines, the most appealing alternatives to cisplatin are probably the *trans*-Pt(IV) and dinuclear bisplatinum complexes. Although these compounds are only in preclinical development they provide a structural alternative to the *cis*-Pt(II) mononuclear derivatives which could strongly affect the cellular processing of the platinum-DNA adducts. Such a difference in mechanism might provide a means of circumventing the current limitations of cisplatin and carboplatin chemotherapy.

3
Interaction of Cisplatin and Related Complexes with DNA

3.1
Kinetics and Order of Cisplatin Binding to Adjacent Purines in DNA

Cisplatin is hydrolyzed in aqueous solution to give first the monosubstituted $[Pt(NH_3)_2(H_2O)Cl]^+$ and eventually the diaqua $[Pt(NH_3)_2(H_2O)_2]^{2+}$ species. The first hydrolysis reaction occurs with a $t_{1/2}$ of about 2 h [50–52] and activates cisplatin for DNA binding. The monofunctional aqua intermediate binds to the N7 atom of the guanine residue through displacement of water and not chloride [51].

The reaction of $[Pt(NH_3)_2(H_2O)Cl]^+$ with chicken erythrocyte DNA was followed by ^{195}Pt NMR spectroscopy by using isotopically enriched ^{195}Pt cisplatin [51]. This study clearly demonstrated that formation of the 1,2-*intra*strand $[Pt(NH_3)_2\{d(GpG-N7(1),N7(2))\}]$ cross-link occurs through two successive pseudo-first-order reactions. The first is formation of the monofunctional adduct $[Pt(NH_3)_2Cl(N7-pGp)]$, which proceeds with same rate constant as hydrolysis of the first chloride ion from cisplatin ($10.2\pm0.7\times10^{-5}$ s^{-1} vs. $9.5\pm1.3\times10^{-5}$ s^{-1} at 37 °C and pH 6.5). The second step is closure of the $[Pt(NH_3)_2\{d(GpG-N7(1),N7(2))\}]$ chelate with a rate constant of $9.2\pm1.4\times10^{-5}$ ($t_{1/2}=2.1\pm0.3$ h). A stepwise mechanism is illustrated in Scheme 1.

In this work it was determined that the kinetic parameters for *trans*-diamminedichloroplatinum(II) binding to DNA are very similar to those of cisplatin. The formation of monofunctional adducts is nearly identical (k=9.6×10^{-5} s^{-1}, $t_{1/2}=2.0\pm0.1$ h, for *trans*-DDP), whereas the second step to form the bifunctional adduct is slightly slower (k=6.3×10^{-5} s^{-1}, $t_{1/2}=3.1\pm0.1$ h) for the *trans*analog. This result is interesting considering that *trans*-DDP does not form a 1,2-*intra*strand chelate, but instead produces 1,3-*intra*- and *inter*strand adducts. This experiment shows that it is not the differential rate of closure of the mono- to bifunctional adducts that is responsible for the biological inactivity of *trans*-DDP, as had been previously proposed [53].

It had been suggested [51] that in vivo the positively charged $[Pt(NH_3)_2(H_2O)Cl]^+$ ion diffuses to the polyanionic DNA and then rapidly migrates along the helix to the most favorable GpG binding sites. This proposal was subsequently evaluated [54] through the use of the cisplatin analog *cis*-$[Pt(NH_3)(NH_2C_6H_{11})Cl(H_2O)]^+$, a metabolite of the orally active platinum drug JM221 [9] (Fig. 5). An HPLC analysis showed that $[Pt(NH_3)(NH_2C_6H_{11})Cl(H_2O)]^+$ binds to unplatinated d(GpG)-containing oligonucleotides at a rate that depends linearly on the concentration of the platinum drug [54]. In addition, the binding of the platinum compound to either the GpG-containing T_7GGT_7 or the phosphorothioate-containing $T_8p(S)T_8$ hexadecanucleotides occurred at a rate 35–40 times faster than to the corresponding GpG or Tp(S)T dinucleotides. Although the Pt-GpG interaction occurs on the DNA bases and the Pt-pS adduct on the helix backbone, a similar rate enhancement was observed in each case. This result

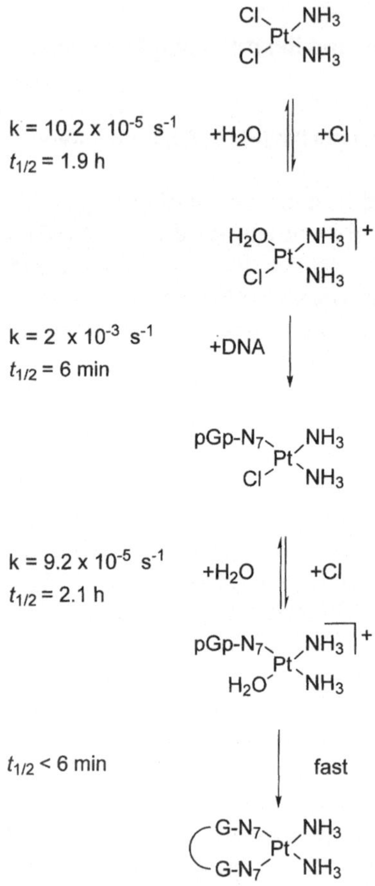

$$\underset{Cl}{\overset{Cl}{\diagdown}}\underset{NH_3}{\overset{NH_3}{\diagup}}Pt$$

$$k = 10.2 \times 10^{-5}\ s^{-1} \qquad +H_2O \parallel +Cl$$
$$t_{1/2} = 1.9\ h$$

$$\left[\underset{Cl}{\overset{H_2O}{\diagdown}}\underset{NH_3}{\overset{NH_3}{\diagup}}Pt\right]^{+}$$

$$k = 2 \times 10^{-3}\ s^{-1} \qquad +DNA$$
$$t_{1/2} = 6\ min$$

$$\underset{Cl}{\overset{pGp\text{-}N_7}{\diagdown}}\underset{NH_3}{\overset{NH_3}{\diagup}}Pt$$

$$k = 9.2 \times 10^{-5}\ s^{-1} \qquad +H_2O \parallel +Cl$$
$$t_{1/2} = 2.1\ h$$

$$\left[\underset{H_2O}{\overset{pGp\text{-}N_7}{\diagdown}}\underset{NH_3}{\overset{NH_3}{\diagup}}Pt\right]^{+}$$

$$t_{1/2} < 6\ min \qquad\qquad fast$$

$$\underset{G\text{-}N_7}{\overset{G\text{-}N_7}{\diagdown}}\underset{NH_3}{\overset{NH_3}{\diagup}}Pt$$

Scheme 1

indicates that weak metal ion-DNA surface interactions can affect the rate of DNA platination independent of the type of adduct. It is probable that such weak interactions play an important role in determining the effectiveness of DNA-binding drugs and affect structure-function relationships [55,56].

Recently, the pathways of adduct formation of activated cisplatin on both single- and double-stranded oligonucleotides was investigated by ^{1}H and ^{15}N NMR spectroscopy [57,58] by using the 14-mer duplex d(ATACATGGTACATA) · d(TAT GTACCATGTAT). The results of this study indicate that the reaction of cisplatin with either single- or double-stranded DNA proceeds through the monoaqua species [Pt(NH$_3$)$_2$(H$_2$O)Cl]$^+$, as had previously been reported [51], and that the reaction depends on the rate of hydrolysis of the first chloride ion.

The use of ^{15}N NMR spectroscopy allowed the two different ammine ligands of the cisplatin to be distinguished and provided evidence for two different monofunctional adducts on both the GG-containing strand and the full length 14-mer duplex [58]. The authors were not able to determine whether the faster forming monofunctional adduct was the same in both ss and ds DNA, however. No NMR signals due to $[Pt(NH_3)_2(H_2O)(N7\text{-}pGp)]$ were observed arising from the hydrolysis product of the monofunctional adduct during the course of this study, indicating that chelate formation is rapid relative to the chloride hydrolysis rate, as previously reported [51]. In both single-stranded and duplex DNA, one of the guanine residues is platinated four times faster than the other, and the monofunctional adduct that forms faster on the duplex also forms the bifunctional adduct more rapidly. On dsDNA the ring-closure rates between the two monofunctional adducts differed by an order of magnitude.

The authors proposed that the more stable monofunctional adduct obtained with duplex DNA involves the 5'-guanine residue. This proposal is based on recent structural work for an 8-mer duplex that reveals greater strain in the *intra*strand $[Pt(NH_3)_2\{d(GpG\text{-}N7(1),N7(2))\}]$ chelate occurs on the 3' side, details of which are discussed in Sect. 4. It has been suggested that attack at the 3'-guanine residue would cause distortion of the duplex and encourage ring closure toward the 5'-G [58]. No direct evidence is provided in this study, however, for distortion of the helix structure upon monofunctional adduct formation at the 3'-guanine. In addition, previous studies of cisplatin analogs have indicated that monofunctional adducts do not distort the helical structure of the duplex [18]. It is possible that the normal structural features of the B-form duplex prevent closure to the bifunctional chelate from the 3' to 5' direction, perhaps by hydrogen bonding to one of the ammines and by inhibiting hydrolysis of the second chloride ion, a requirement for chelate formation. In the idealized B-form duplex, the distance from a monofunctional Pt adduct to a 5'-(n-1) purine N7 atom is ~3 Å, whereas the distance to a 3'-(n+1) purine N7 is almost 5 Å [59].

This difference may explain why the cisplatin-GpA adduct is not observed. The most electron-rich sites for platinum site are the N7 atoms of guanine. If the adenine N7 monofunctional adduct were formed first, the strongly basic guanine chelate could promote hydrolysis and binding. The observation of a long-lived intermediate, however, may indicate that initial attack at the guanine of a GpA site prevents hydrolysis and closure which is not overcome by the less basic adenine.

3.2
Sequence Specificity of Cisplatin and Carboplatin In Vitro and In Vivo

Carboplatin affords a less toxic cisplatin derivative, the action mechanism of which involves interaction with DNA in a similar manner. In vitro studies of carboplatin have indicated that the distribution of DNA adducts formed upon treatment with this platinum drug are similar to that observed for cisplatin. The extent of DNA damage and specific adducts formed were monitored by the conformational changes induced in pUC18 plasmid DNA using agarose gel electrophoresis [60]. Platination of the superhelical plasmid unwinds the duplex and winds the negative superhelical structure. The extent of superhelical winding can be determined through reduced mobility on an agarose gel [61]. The conformational changes were dose-dependent up to a 0.1 mole ratio of cisplatin to nucleotide following incubation for 16 h at 37 °C. The same experiment with carboplatin showed no reduced mobility up to a 0.5 mole ratio for 16 h at 37 °C. Longer incubation times up to 120 h and mole ratios of 0.9 to 1.0 were required to effect the maximum reduced mobility observed with cisplatin. These results indicated the rate of platination by carboplatin to be significantly slower than by cisplatin, a difference in kinetics due primarily to the hydrolysis rate of the leaving groups on the two compounds.

Another study [62] used atomic absorption spectroscopy to determine the level of platination of salmon sperm DNA(ssDNA). Again the rate and extent of platination differed dramatically for carboplatin versus cisplatin. Carboplatin required a dosage level 230-fold higher than that necessary with cisplatin after a 4 h incubation period at 37 °C. The extent of platination increased linearly with time for up to 120 h, indicating a much slower reaction rate with carboplatin.

This study also examined the types of platinum adducts formed upon reaction with carboplatin. Not surprisingly, the major adduct formed (58%), as detected by an enzyme-linked immunoabsorbent assay, was the 1,2-*intra*strand cross-link, $[Pt(NH_3)_2\{d(pGpG)\}]$. This result is very similar to that observed for cisplatin under similar conditions [10]. The other minor products formed were $[Pt(NH_3)_2\{d(pApG)\}]$, the *inter*strand dG-Pt-dG, and monofunctional Pt-dG detected at levels similar to those found in cisplatin experiments with values of 11, 9 and 22%, respectively. This distribution of products is expected given that the active forms of both drugs are identical.

A study with cultured Chinese hamster ovary (CHO) cells revealed that a carboplatin dose level of 0.70 mM was required to obtain 10% survival level after a 1 h exposure, whereas 40 µM cisplatin can achieve the same level of survival [62]. This study showed that the relative efficiency of platination is more similar in vivo than in vitro, but that a nearly 20-fold increase in dose level was still required for carboplatin. The slower rate of hydrolysis is probably responsible for the lower nephrotoxicity observed for carboplatin. Whereas the CHO study revealed the expected distribution of adducts for cisplatin, similar to that observed for ssDNA, carboplatin showed a very different result. The major adducts included a larger percentage of dG-Pt-dG (33–41%), a much lower level of $[Pt(NH_3)_2\{d(pGpG)\}]$ (28–33%) species, and ~16% $[Pt(NH_3)_2\{d(pApG)\}]$ [62]. This difference may be due to the relative reaction rates of carboplatin and cisplatin, allowing for intracellular processes to modulate the activity of the more slowly reacting species. The interaction of histones with pGpG sites on the DNA was suggested by the authors as a possible reason. In another report, again using CHO cells, it was suggested that, due to the relative distribution of Pt adducts in this cell line, the $[Pt(NH_3)_2\{d(ApG)\}]$ adduct may be the cytotoxic lesion since it is formed at similar levels and dose efficacies with both cisplatin and carboplatin [63].

3.3
DNA Targets for Cisplatin May Reflect Its Ease of Removal (Repair)

Mitochondrial DNA (mtDNA) might be a good target for cisplatin and other DNA-damaging agents because it lacks histones [64]. Most studies of cisplatin-DNA adducts have focused on their incorporation into genomic DNA (gDNA). An investigation [65] of the relative distribution of platinum adducts in CHO cells was undertaken by dissociation-enhanced lanthanide fluoroimmunoassay (DELFIA), which uses an antiserum elicited against cisplatin-modified DNA [65] and immunoelectron microscopy. The results with either method showed that cisplatin is incorporated at a level sixfold higher in mtDNA than in gDNA. The higher levels of cisplatin in mtDNA were correlated with both an initial increase in binding and a reduced level of repair of the resulting DNA adducts [66]. These observations are interesting since mitochondria are deficient in nucleotide excision repair, a means by which cisplatin adducts are removed from gDNA, providing another possible mechanism for acquired cisplatin resistance. Possibly the effectiveness of cisplatin could depend on binding to specialized types of DNA, such as mtDNA or telomeric regions of the chromosomes.

3.4
Cisplatin Interaction with Telomeres and Telomerase

Telomeres are long tandem repeats of DNA which occur at the ends of chromosomes and contain a repeating sequence unit of 5-TTAGGG-3' in humans. This guanosine-rich region of the chromosome is potentially a good target for cisplatin, and if adducts are formed they could block replication of the telomeric region of the chromosome. Telomeres protect the chromosome from degradation, prevent loss of genomic information, and may provide a means of maintaining nuclear structures [67]. They are responsible for immortalization of cells and are lost at a rate of 50–200 bp per cell division. A ribonucleoprotein, known as telomerase, containing a RNA template, synthesizes telomeric repeats during active cell division. Telomerase activity may be responsible for the unrestricted growth of malignant tumors [68].

In a recent study, it was demonstrated that cisplatin can shorten telomeres in HeLa cells in culture [69]. At high levels of exposure (5 µM), DNA replication was inhibited and the cells were killed before being able to divide. At lower levels of cisplatin incubation (0.5 µM), however, shortened telomeres were observed through the use of a Southern assay. The telomere loss in these cells was >6 kbp, which apparently is a sufficiently lethal effect to cause non-apoptotic cell death. This selective telomere damage at very low cisplatin concentrations indicates that the telomeric G-rich sites are good targets for cisplatin. It is possible that cisplatin adducts on telomeres contribute significantly to the anticancer activity of the drug [69].

4
Structural Studies of Platinum-DNA Complexes

Early studies of cisplatin binding to nucleobases and nucleotides indicated that significant distortion of the DNA structure occurred upon platination of purine-rich sites. Whereas the distortions observed in the crystal structures of [Pt(NH$_3$)$_2$\{d(pGpG)\}] [12,45] and [Pt(NH$_3$)$_2$\{d(CpGpG)\}] [14] gave evidence of the structural changes that could occur upon cisplatin adduct formation, it was of interest to examine the structures of duplex DNA containing a discrete 1,2-*intra*strand adduct. Recently, a number of high-resolution structure determinations have appeared for the major cisplatin adduct on duplex DNA [47–49,70] as well as minor adducts [71,72].

4.1
Solution and Crystal Structures of *Intra*strand Cross-Linked Duplex DNAs

4.1.1
Crystal Structure of a Dodecamer Duplex Containing the Major Cisplatin 1,2-GpG Intrastrand Adduct

The first crystal structure of a site-specifically platinated duplex was reported [46,70] at 2.6 Å for the DNA dodecamer d(C$_1$C$_2$U^{Br}$_3$C$_4$T$_5$G$_6$*G$_7$*T$_8$ C$_9$T$_{10}$C$_{11}$C$_{12}$) · d(G$_{13}$G$_{14}$A$_{15}$G$_{16}$A$_{17}$C$_{18}$C$_{19}$A$_{20}$G$_{21}$A$_{22}$G$_{23}$G$_{24}$) (S1), where G*G* indicates the site of platination and U^{Br} is a 5-bromo-U heavy-atom derivative of the native DNA sequence used for phasing. Crystals of this duplex were obtained by separately preparing and purifying both the platinated and unmodified oligonucleotide strands and then annealing them to give the duplex containing a single cisplatin adduct. The structure was solved by multiple isomorphous replacement (MIR) utilizing three different brominated heavy-atom derivatives in addition to the native duplex DNA. The triclinic crystals contained two independent molecules which allowed two different but very similar structures of the platinum-modified duplex (molecules A and B) to be obtained.

The X-ray structure of the cisplatin-modified dodecamer duplex revealed many interesting features. First, the duplexes are bent towards the major groove with the bend locus near the site of platination. Figure 6A shows a MOLSCRIPT [73] drawing of molecule A from this analysis [70]. The structure revealed two different bend angles, one of ~39°, similar to what had been previously reported from gel electrophoresis studies [17,18], and the other ~55°. The bend is due in large part to a roll of 26° towards the major groove of the G*G* pair upon platination. This dihedral angle between the guanine base planes is quite shallow when compared to that observed in the crystal structure of the single-stranded dinucleotide complex of 87°. The duplex formation and crystal packing (discussed below) preclude more complete destacking of the bases. Base pairing is retained throughout much of the helix and causes significant distortion of square-planar platinum geometry. The Pt atom lies ~1 Å out of the plane of the

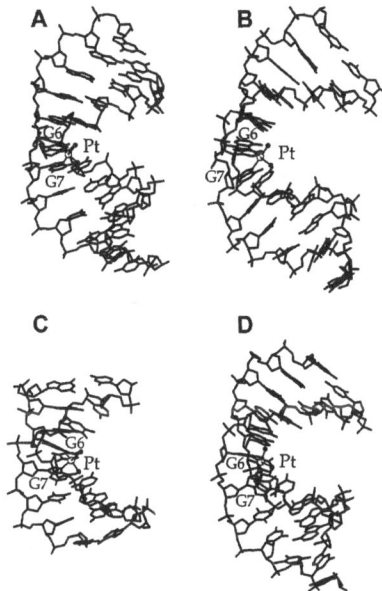

Fig. 6. MOLSCRIPT [73] representations of the recent structure determinations of cisplatin-1,2-*intra*strand DNA adducts. **A** X-ray crystal structure of a dodecamer duplex S1, redrawn from coordinates in [70]; **B** NMR-derived solution structure of the same duplex, S2; **C** solution structure of an octameric duplex DNA, S3, redrawn from coordinates in [47]; **D** solution structure of an undecamer duplex containing the site-specific platinum adduct determined by the addition of long-range distance constraints, S4

guanine rings, indicating that the induced structure causes significant strain on the Pt-N7 bonds.

The base pairs of this duplex are only minimally disrupted even through the region of the platinum adduct. The four base pair region around the cisplatin adduct for four recent structure determinations [47–49,70] is shown in Fig. 7. In Fig. 7A it is clear that the binding of *cis*-[Pt(NH$_3$)$_2$] to the adjacent N7 atoms of G6* and G7* causes only minor propeller twisting (ω) of the base pairs G6*–C19 and G7*–C18, with a larger distortion occurring at the G6*–C19 bp in molecule A ($\omega=-22°$) and G7*–C18 bp in molecule B ($\omega=-19°$). It has been suggested [70] that the shallow roll caused by the platinum adduct may facilitate the hydrophobic helix interactions which lead to a novel A-form–B-form DNA fusion.

This A-B fusion, which occurs in the crystal structure of the platinated duplex S1, is most likely due to crystal packing. The first 8 bp to the 5'-side of the platinum adduct, referenced to the platinum-modified strand, exhibit A-form DNA parameters. The sugar puckers are in the N conformation (C3'-endo), the phosphate-phosphate distances have an A-type distance of 5.8 Å, and the DNA bases are displaced from the helix axis. In addition the minor groove is widened (9.2–11.2 vs 5.7 Å for B-form DNA) and flattened over the first eight base pairs. The A-form end of the DNA also packs in the crystal in a conventional A-type manner. The 5'-end of the helix packs into the minor groove of another duplex. In this case the C1–G24 base pair contacts a second duplex at the G7–T8 base step. This contact is responsible for extending A-type parameters to the 5'-end of the duplex [70].

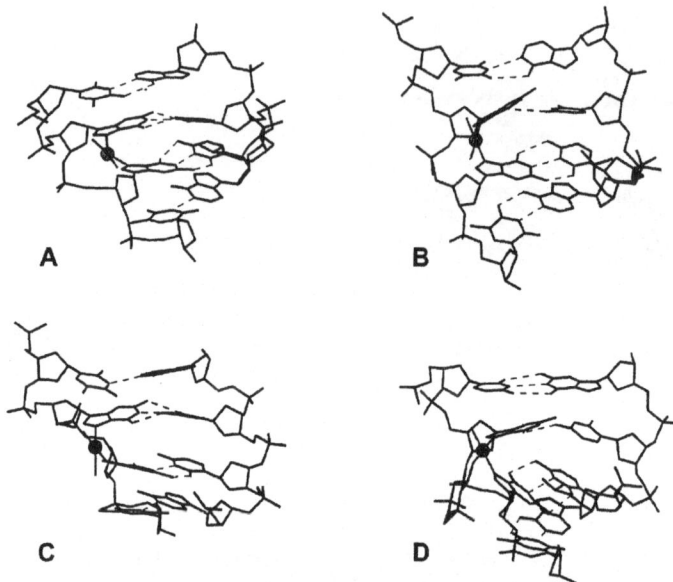

Fig. 7. Central four base pairs of the cisplatin-modified DNA structures A) S1; B) S2; C) S3, D) S4, shown in Fig. 6. Intact base pair hydrogen bonds are indicated with *dashed lines*

From base pair C9–G16 through C12–G13 the helix resembles B-form DNA, having a longer average phosphate-phosphate distance of 6.6 Å and S sugar conformations (C2'-endo). This B-form end of the A-B fusion also exhibits a common B-DNA crystal packing motif. The 3'-end of one helix (base pair C12–G13) stacks against the 3'-end of a second duplex, effectively creating a continuous B-form duplex. These intermolecular contacts maintain the A-B fusion observed in these duplexes. Some of the observed structural parameters, such as the opened minor groove, would, if preserved in solution, explain how cellular proteins might recognize and process cisplatin-DNA adducts (see below).

4.1.2
Solution Structure of a Dodecamer Duplex Containing the Major Cisplatin 1,2-GpG Intrastrand Adduct

In order to address whether some of the unique structural features found in the crystal structure of cisplatin bound to duplex DNA might arise from crystal packing, the solution structure of a cisplatin-modified dodecamer containing the same DNA sequence was determined by NMR spectroscopy [48]. The solution structure of the DNA dodecamer $d(C_1C_2T_3C_4T_5G_6{}^*G_7{}^*T_8C_9T_{10}C_{11}C_{12}) \cdot d(G_{13}G_{14}A_{15}G_{16}A_{17}C_{18}C_{19}A_{20}G_{21}A_{22}G_{23}G_{24})$, S2, was determined by combined 2D ^{1}H-^{1}H NMR spectra and restrained molecular dynamics (rMD). A model of the final structure is

Table 1. Selected structural parameters of cisplatin/DNA structures

Complex	S1 [46,70]	S2 [48]	S3 [47]	S4 [49]
DNA form	A/B junction	primarily B	primarily B	primarily B
Minor groove width/depth (Å)	9.5–11.0/3.0	9.4–12.5/1.4	4.5–7.8/3.2	9.0–12/2.1
Average P-P distance (Å)	5.5	6.9	6.8	6.8
Roll at platinated bases (°)	26	49	42	59
Pt atom displacement from guanine ring planes	1.3 Å, 5' 0.8 Å, 3'	0.8 Å, 5' 0.8 Å, 3'	1.0 Å, 5' 0.8 Å, 3'	0.5 Å, 5' 0.65 Å, 3'
Average helical twist (°)	32	25	25	26
DNA bend (°)	39 and 55[a]	78	58	~81

a Two independent molecules.

shown in Fig. 6B. In solution some of the features of the crystal structure are exaggerated, probably due to the absence of crystal packing forces. In particular, the G6*G7* roll and the helix bend are larger, i.e. 49° and 78°, respectively. The greater helix deformation may have been prevented in the crystal structure due to helix-helix packing interactions at the site of platination. The larger roll between guanine bases relieves the strain imposed on the platinum-guanine-N7 atom bonds to some extent, but even in solution, the Pt atom is displaced from each base plane by ~0.8 Å. Structural parameters are compiled in Table 1.

NMR spectra in H_2O revealed 11 of 12 imino proton resonances, indicating the completeness of duplex formation in this structure. Only the imino proton from G6* exchanged with solvent on the NMR time scale, implying a disruption of base pairing only at this 5'-guanine. This disruption is represented in Fig. 7B. In this base pair only the G6*O4–C19N4 hydrogen bond remains due to a large propeller twist of ω=–20°, nearly identical to that found in the crystal structure. Duplex formation prevents complete destacking of the guanine bases that is observed in the crystal and solution structures of the Pt-GpG complex, the structural effects of the platinum adduct being spread over many base pairs in the duplex causing a larger global deformation. This type of deformation, resulting in a disruption of the DNA helix centered at the platinum lesion, was previously proposed based on the reaction of antinucleoside antibodies to cisplatin-1,2-*intra*strand adducts [74].

Although considerably distorted from B-form DNA, duplex S2 has predominately B-type sugar puckers (S conformation) and internucleotide phosphate-phosphate distances (6.9 Å) in solution. Only the 5'-platinated guanosine (G6*) deoxyribose ring occurs in the C3'-endo conformation as suggested by the 2D-PECOSY and confirmed in the 2D-NOESY spectra. The remaining 23 ring sugar puckers exist in the C2'-endo or closely related C3'-exo conformation. Whereas these sugar-phosphate backbone parameters are quite different from those observed in the crystal structure of this same duplex, a remarkable similarity exists in the minor groove deformation caused by the cisplatin adduct.

The widened and flattened minor groove observed in the X-ray structure, which is reminiscent of A-form DNA, is observed in solution as well. The cisplatin lesion causes an unwinding of ~25° at the site of platination, which extends the minor groove widening opposite the adduct over five base pairs (C4–G21 to T8–A17). In this region of the duplex, the minor groove width varies from 9–12 Å, and the average groove depth is 1.4 Å; B-form DNA parameters are 5.7 and 7.5 Å, respectively [48]. Superposition of the first eight base pairs of the crystal and solution structures of this duplex provides a graphical comparison of the similar global deformation. Figure 8 displays a view of this superposition in the minor groove. Although the overall RMSD for these eight base pairs is 2.7 Å, the sugar-phosphate backbones, represented by ribbon diagrams, reveal that a similar profile is presented in the minor groove upon formation of the cisplatin adduct.

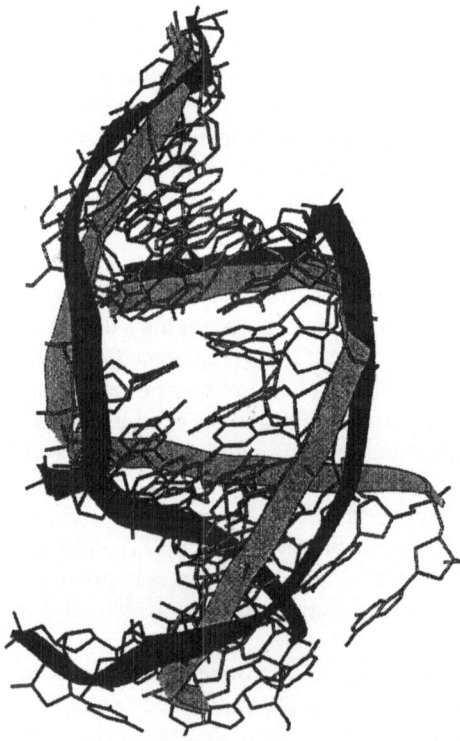

Fig. 8. Superposition of the top eight base pairs of the X-ray crystal (*gray*) and NMR solution (*black*) structures of the duplex d(CCTCTG*G*TCTCC)· d(GGAGACCAGAGG) showing the similarity of the minor groove widths. Reproduced with permission from [48]

4.1.3
Solution Structures of Cisplatin Cross-Linked Octamer, Undecamer, and Palindromic Duplexes

The solution structures of three other duplex DNAs containing the major cisplatin adduct have been determined since the first report of crystal structure S1 [46]. One is an octamer duplex having the same internal sequence d(-TG*G*T-) as that in S1 [47]. The second employed a spin-labelled platinum complex on an undecamer duplex and was solved by using both conventional NMR methods and long-range proton-electron distance constraints [49]. The third was a palindromic dodecamer used to study the effects of two adducts on a single duplex [75]. The details of these structures are described below.

Solution structure of a duplex octamer containing the cisplatin 1,2-intrastrand cross-link. The structure of an octamer duplex, d($C_1C_2T_3G_4$*G_5*$T_6C_7C_8$) · d($G_9G_{10}A_{11}$ $C_{12}C_{13}A_{14}G_{15}G_{16}$), S3, was determined by using conventional 2D NMR spectroscopy and molecular dynamics [47]. Its general structural motif was similar to that in S1. During the course of this study, however, the authors found the short duplex to be rather unstable in the relatively low salt conditions of their NMR buffer (150 mM NaCl). It slowly converted to a new species (vide infra). The intrastrand cross-link formed on this duplex causes a roll of 42° of the guanine bases towards the major groove, leading to a large helix kink angle of 58°. A model of this structure is shown in Fig. 6C. Although S3 shows the extended bend angle found for S2 versus S1, it was not accompanied by the exaggerated widening of the minor groove that occurs in both S1 and S2 [48]. This reduced effect is possibly due in part to the limited helicity of the shorter duplex. The bound platinum adduct causes a rather large duplex unwinding (~21°) and the minor groove is widened to about 7.4 Å at the d(G4pG5) base step. Disruption of the B-form minor groove is limited to those two base pairs, however.

Solution structure determination of a duplex undecamer containing the cisplatin 1,2-intrastrand cross-link through the addition of long-range distance constraints. A limitation that has existed in structure determinations of DNA by NMR NOESY methods has been the small through-space range of the nuclear Overhauser effect, ≤5 Å. Although interproton distances of this magnitude allow the local geometry of DNA helices to be determined to high resolution, it does not give direct experimental evidence for global features such as bend angles or long range unwinding. Recently this issue has been addressed in a platinated-DNA solution structure determination by use of a spin-labeled platinum compound, which allowed the acquisition of a number of long-range electron-proton distance constraints for use in restrained molecular dynamics [76].

Cisplatin analogs containing a cyclohexylamine ligand in place of one of the ammines bind to the GpG site in DNA in a manner similar to cisplatin [9,54]. The reaction of the spin-labeled compound [Pt(NH_3)(4-amino-TEMPO)ClI] with the undecamer d(CTCTCGGTCTC) yields two platinated DNA isomers, one of which (3'-TEMPO) was isolated and annealed to its complementary strand to give a spin-labeled duplex undecamer. The structure of the reduced diamagnetic duplex $d(C_1T_2C_3T_4C_5G_6{}^*G_7{}^*T_8C_9T_{10}C_{11}) \cdot d(G_{12}A_{13}G_{14}A_{15}C_{16}C_{17}G_{18}A_{19}G_{20}A_{21} G_{22})$, S4, was determined by using conventional 2D NMR methods and distance-restrained molecular dynamics [49]. This structure was further refined through the addition of 99 additional long-range (<20 Å) electron-proton distance constraints. The electron-proton distances were obtained through quantitation of the distance-dependent relaxation of resolved proton resonances. The local geometry of the DNA duplex in the diamagnetic structure resembles that of both the X-ray and solution structures of the dodecamer duplex described above. The sugar puckers for this duplex are primarily in the S-conformation. Only G6*, the 5'-platinated guanosine and the terminal cytosine C11 are in the N conformation, and the average phosphate-phosphate distance is ~6.7 Å. These parameters agree well with those observed in the dodecamer solution structure. In addition, the platinum lesion causes a roll of ~50° at the G6*pG7* step, distorting the 5'-G-C base pair. The minor groove in this structure is rather distorted from that of B-form DNA, with a groove width range of 9–12 Å over the four base pairs T4–A19 to G7–C16. This minor groove deformation is similar in scale but smaller in extent (4 vs. 5 base pairs) than that observed for S2. The structure of S4 is depicted in Fig. 6D and details of the central four base pairs are shown in Fig. 7D.

The addition of the long-range distance constraints during the structure refinement of the platinated undecamer duplex resulted in changes from the diamagnetic structure primarily at the 5'-end of the duplex. In the final refined structure the base pairs T2–A21, C3–G20, and T4–A19 were the most affected; in all three cases causing the bases to bend further towards the major groove. The resulting helix bend angle of ~81° is even larger than that observed for the dodecamer in solution. The geometric changes caused by addition of the long-range electron-proton distance constraints resulted in a structure which more closely resembled the X-ray structure of the dodecamer duplex than the structures of either the dodecamer or octamer duplex. While the superposition of all of the heteroatoms for the first eight base pairs of the solution and crystal structures of the dodecamer provided a good structural comparison with an RMSD of 2.7 Å, the superposition of the sugar-phosphate backbone of the undecamer and the corresponding 11 base pairs of the crystal structure gave an RMSD of 2.0 Å. This value may be compared to backbone-backbone RMSD values of 2.3 Å for the diamagnetic structure, 2.8 Å for the octamer, 3.8 Å for the dodecamer and 2.6 Å for the first eight base pairs of the dodecamer.

Based on the analysis of the final refined structure of the spin-labeled duplex, it appears that the use of the paramagnetic constraints is especially important in defining the ends of duplex structures. Typically these regions of the DNA are poorly determined due both to some intramolecular motion and a lack of the full complement of interproton NOEs for these base pairs. The use of the long-range distance constraints allows a better definition of the global parameters of the duplexes, and in this case gives a structure having a very similar global structure to that obtained in the solid state.

Solution structure of a palindromic duplex dodecamer containing two cisplatin 1,2-intrastrand cross-links. A study using palindromic DNA sequences having adjacent guanosines was conducted in order to examine the effect of the cisplatin adduct on self-complementary duplex DNA [75]. The duplex d(GACCATATG*G*TC) was designed so that the [Pt(NH$_3$)$_2${d(GpG)}] adducts would be positioned 180° apart in B-form DNA. Each of the cisplatin adducts causes a local helix bend of ~40°. The two lesions together cause an effective 13Å dislocation of the helix axis. The authors do not describe this structure in sufficient detail to reveal any of the more important structural features discussed above. It appears that the minor groove remains undistorted between the platinum adducts and that the roll angle is rather shallow. It is suggested that these types of structures, having cisplatin adducts close to the end of the duplex, may lead to the instability that has been observed for the 1,2-intrastrand adduct in the octamer duplex previously described [47].

To address some of the issues regarding the location of the cisplatin adduct in duplex DNA, the self-complementary duplexes d([c^{7}A$_1$]C$_2$C$_3$[c^{7}G$_4$][c^{7}G$_5$]C$_6$C$_7$ G*$_8$G*$_9$T$_{10}$) and d([c^{7}G$_1$]C$_2$C$_3$[c^{7}G$_4$]C$_5$G*$_6$G*$_7$C$_8$) were prepared. These oligonucleotides contain 7-deaza-purines, which prevent platination. NMR studies on these duplexes indicate that very different structures exist for these three platinated oligomers. Whereas a single 1,2-intrastrand adduct is formed on d([c^{7}A] CC[c^{7}G][c^{7}G]CCG*G*T) and the platinated guanosines contain the commonly observed G*pG*–C3'-endo–C2'-endo sugar puckers, the 5'-platinated guanosine has the unusual syn-conformation commonly found in single-stranded DNA. In addition, imino protons were observed only for central nucleotides G4 and G5, indicating limited base pairing beyond the core of this complex. These data suggest that the platinated end of the DNA exists primarily in a single-stranded configuration. The platinated oligomer d([c^{7}G]CC[c^{7}G]CG*G*C) would have the two 1,2-intrastrand adducts separated by only two base pairs in an intact duplex. The NMR data indicate that two different species exist initially which can be converted to single species containing numerous Watson-Crick base pairs.

These self-complementary complexes demonstrate that cisplatin adducts at the ends of duplexes can lead to a number of varied structures. Such end adducts, as well as the adduct found in the center of a short octamer duplex, are unstable and can convert in the presence of chloride ion to interstrand adducts [47,75]. When a single-stranded end is formed, these oligomers are not Cl⁻ sensitive, however [75]. In none of the longer duplexes that have been structurally characterized [48,49,70] has such instability of the platinum lesion been observed.

4.2
Solution Structures of Interstrand Cross-Linked Duplex DNAs

The interstrand cross-link (ICL) represents a small fraction (~5%) of the adducts formed by cisplatin on duplex DNA [10]. Although the 1,2-intrastrand cross-link is generally believed to represent the cytotoxic lesion, the ICL [1] may

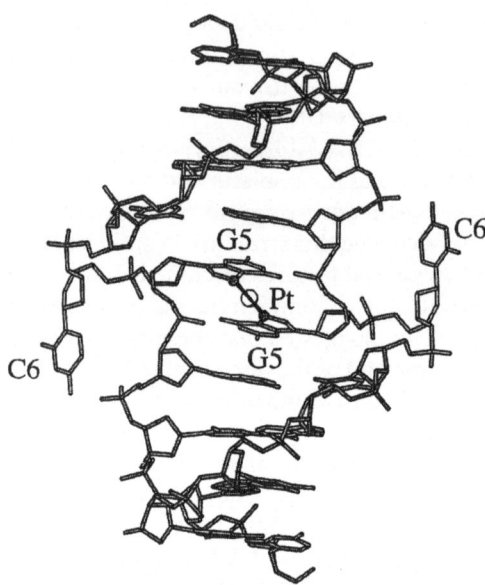

Fig. 9. MOLSCRIPT drawing of the cisplatin-*inter*strand cross-linked decamer structure solved by NMR spectroscopy and rMD methods. Redrawn from coordinates given in [71]

also contribute, especially since the 1,2-*intra*strand cross-link can convert to this form [47,75] (see Sect. 4.4). Formation of the cisplatin ICL occurs best at a GpC step [77] and is expected to cause significant distortion of B-form DNA due to the long (>7 Å) distance between N7 atoms of the guanine bases involved. Gel electrophoresis measurements indicated that ICL adducts cause a bend of 45° in the DNA duplex, unwinding it by ~80° [78,79].

The solution structures of two different duplexes containing the cisplatin *inter*strand cross-link have been determined by NMR and rMD [71,72], one of which, d(CATAG*CTATG)$_2$, is shown in Fig. 9 [71]. The large distortion caused by the ICL results in a inversion of the duplex allowing the G(N7)-platinum-G(N7) adduct to become placed in the minor groove. The guanine bases can shift their orientation by inversion of the deoxyribose ring resulting in its O4' atom being pointed in a direction opposite to the others in the helix. This distortion is further accommodated by the displacement of the C6 cytosines which are extra-helical in this structure (Fig. 9). Extensive unwinding caused by the ICL adduct produces a local left-handed helix geometry about the platinum atom resembling that of Z-DNA [80].

The combination of these structural deformations leads to a helix bend of ~40° towards the minor groove, primarily the result of a G*G* dihedral angle of 38°, and an unwinding angle of 76° [72], values close to those obtained by gel electrophoresis. Again, in this very different structure, a shallow roll angle between the platinated guanine bases displaces the Pt atom by 0.6 Å [71,72] and presumably strains the Pt–N7 bonds. Interestingly, in both structures containing the cisplatin ICL, the platinum geometry is distorted. In one case [71], the platinum atom maintains formal square-planar geometry, but with a pseudo-octahedral configuration owing to the orientation of the G5* phosphate oxygen atom positioned at an axial position, the Pt-O(P) distance being 3.2 Å. This electrostatic interaction may help to stabilize the distorted helical geometry in these structures. In the second structure [72] a longer Pt-O(P) interaction occurs (3.6 Å) and the platinum geometry is quite distorted from square-planar (N7-Pt-N7 angle, 83°). In addition, the ammine ligands appear to create a pseudo-tetrahedral geometry at platinum, an observation that may reflect a limitation in the force-field used for this structure determination [81]. In both cases, some of the forces which may be necessary for stabilizing unusual platinum-DNA structures have been revealed: base stacking, unconventional hydrogen bonding, and electrostatic interactions [71,72].

4.3
Structural Studies of Other Platinum-DNA Adducts

4.3.1
NMR Structure of the 1,3-GTG Intrastrand Cross-Link

The minor product 1,3-*intra*strand GXG cross-link (where X is C or T) is interesting because it forms a much less distorted structure. The effect of the 1,3-*intra*strand cross-link revealed in an early NMR spectral study [82] indicated that the distortion due to the 1,3-adduct was much larger than that found for the 1,2-cross-link [15]. Subsequently, gel electrophoresis experiments revealed that the cisplatin 1,3-cross-link bent the DNA helix by ~35° [18] and unwound it by 23° [19] in the G*TG* adduct. Recently, the 3D structure of a 13-mer duplex, d(CTCTAG*TG*CTCAC) · d(GTGAGCACTAGAG), was determined in solution by NMR spectroscopy and rMD [83]. Figure 10 displays a MOLSCRIPT representation of the solution structure of a DNA duplex containing the *cis*-[Pt(NH$_3$)$_2$ {d(G(N7)TG(N7))}] adduct.

The 1,3-*intra*strand cross-link does not distort the global helical structure to the extent observed in the 1,2-*intra*strand adducts described in Sect. 4.1, but significantly disrupts local base pairing and stacking at the 5'-side of the adduct. The platinum lesion causes a severe roll of the 5'-guanine base (G6) towards the major groove. The large propeller twist of –81° at the G6-step causes a disruption of the hydrogen bonds to C21. The intervening base T7 is no longer base-paired to A20 and stacks under the distorted G6* base, pushing the thymine ring into the minor groove opposite the platinum adduct. This type of structure had previously been predicted for a related 1,3-*intra*strand *trans*-DDP adduct [84]. Interestingly, the 3'-side of the lesion is almost undistorted. The G8–C19 base pair remains intact and undergoes only a slight positive propeller twist of 11°.

The sugar puckers of guanosines G6 and G8 are similar to those observed in the 1,2-*intra*strand structures [47–49], where the 5'-deoxyribose ring is in a C3'-

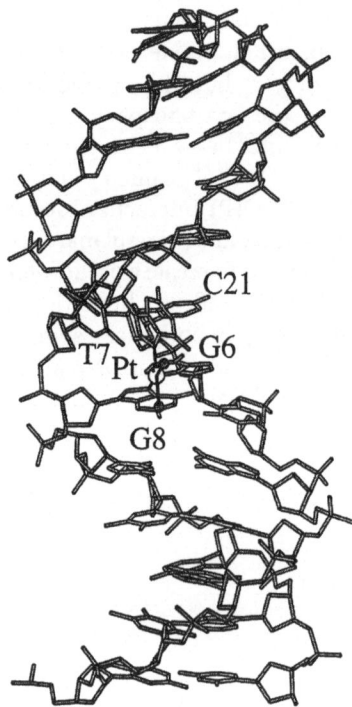

Fig. 10. MOLSCRIPT drawing of the 1,3-*intra*-strand adduct of cisplatin on a 13-mer duplex with the –GTG-bases labeled. Redrawn from coordinates provided in [83]

endo conformation and the 3'-sugar maintains the B-DNA form C2'-endo or S configuration. In this structure the deoxyribose rings of the "complementary" cytosines C19 and C21 are also distorted, having a N/S conformational mixture.

The local distortions induced by the cisplatin 1,3-adduct cause global distortions of the helix. The helix is bent by 24° and unwound by 19°. In contrast to the structural deformation caused by the 1,2-*intra*strand cross-link, the minor groove in this duplex remains in a B-DNA conformation, with an average groove width of 5.8 Å. This narrow minor groove, as well as the presence of the T7 base in the minor groove, may be reasons why HMG1, a protein that shows specific recognition of cisplatin-modified DNA, does not bind to DNA containing the 1,3-*intra*strand adduct [36].

4.3.2
NMR Structure of a Dinuclear Platinum Complex with a Self-Complementary Octamer DNA

The development of dinuclear bifunctional platinum complexes as alternatives to cisplatin was discussed in Sect. 2.4. The structure of one of these complexes, $[\{trans\text{-}PtCl(NH_3)_2\}_2(H_2N(CH_2)_4NH_2)]^{2+}$ (1,1-t,t), bound to the self-complementary oligomer d(CATGCATG), was investigated by NMR spectroscopy and rMD [85,86]. The DNA target of these compounds is presumed to be non-adja-

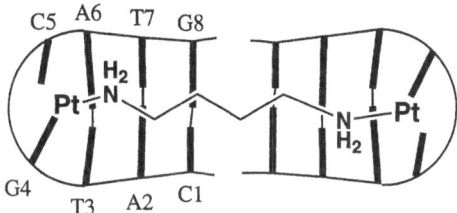

Fig. 11. Schematic representation of the dumb-bell double hairpin structure formed by a bisplatinum compound adduct on a palindromic sequence

cent guanine bases, and the butane derivative was designed to form an *inter*strand adduct in the major groove at a GpC site.

NMR spectroscopy indicated that, upon platination of the DNA, the duplex structure was no longer maintained. Instead, the bisplatinum complex binds to the N7 atoms at the G4 position of two different hairpin DNA structures causing the guanine base to end up in the minor groove. Figure 11 shows schematically how the oligomers are orientated, effectively giving a dumb-bell structure resulting from end-to-end packing of the mini hairpins.

In this novel structure the platinated guanine base is in a *syn* conformation, reminiscent of single-stranded DNA, and the structure of the palindromic sequence d([c⁷A]CC[c⁷G][c⁷G]CCG*G*T) described in Sect. 4.1.3. The *syn* glycosidic angle disrupts the B-form helix and allows the hairpin structure to form by using the internal base pairs of the palindrome. This unique structure formed by the dinuclear platinum complex has been proposed to promote formation of larger distorted DNA structures. The authors have recently suggested that, in longer duplexes, hairpin formation may result in a cruciform structural motif leading to higher order structures [86]. Perhaps such higher order structures could be detected by proteins that recognize and bind to cruciforms and four-way junction containing DNAs [87], such as the HMG-domain proteins discussed in Sect. 5.

4.4
Structural Transitions and Stability of the Cisplatin-DNA Adducts

The structures of the various cisplatin-DNA adducts have revealed that platination can cause a significant distortion from B-form DNA. The formation of cisplatin adducts at the N7 positions of purine bases affords a strong complex which can typically be displaced only by attack by much stronger nucleophiles such as CN⁻ [88]. Recently, however, during the solution structure determination of the octameric duplex S3 containing the 1,2-*intra*strand cisplatin adduct, isomerization to an *inter*strand adduct occurred [47]. The use of 2D NOESY experiments showed the appearance of new peaks corresponding to a different set of amino/imino resonances within 1 d and to a different set of thymine methyl protons within 3 d. The change in the NMR spectrum occurred while the sample was kept at 2 °C. The *intra*strand complex completely converted to this new species over 2 weeks in the NMR tube. A combination of mass spectrometric and 2D NMR spectroscopic analysis revealed cisplatin *inter*strand cross-link formation between the 5'-guanine G4 and the terminal guanine G9 [47,86].

Although *intra*- to *inter*strand isomerization has been reported for DNA adducts of *trans*-DDP [89,90], the foregoing interconversion was unexpected for cisplatin [91]. The isomerization is modulated by Cl$^-$ and results in part from strained Pt–N7 bonds caused by the shallow roll of the coordinated guanine bases [47]. The principal factor in this isomerization, however, may be the use of a short (<1 helix turn) duplex in this structure determination. This conclusion is based on several factors. First, the Cl$^-$ ion concentration in solution of this duplex is 150 mM, similar to that used in the sample buffer of the two NMR structures of a dodecamer and undecamer duplex containing the cisplatin-GpG adduct [48,49]. In both of these latter investigations, 2D NMR spectra collected over a period of 2 to 10 d showed no instability of the cisplatin 1,2-*intra*strand cross-linked products. Furthermore, HPLC and gel electrophoresis analysis of these duplexes revealed only the pure platinated strand and its complement, and no evidence for *inter*strand cross-link formation [92]. Second, the roll of the guanine bases in the structure of this octamer is very similar to that in the crystal structure of the dodecamer duplex [70] (see Table 1). Although this shallow roll may contribute to the destabilization of the Pt-G*pG* adduct, no sign of decomposition occurred during the period of 1 to 12 months required to obtain good crystals of this complex. Finally, the longer duplexes with the additional three or four base pairs reveal that the added length allows a return to a B-form structure at the ends of the helix. In these cases [48,49,70], the widened minor groove closes from 11–12 to 8 Å for the crystal and to 5–6 Å for the solution structures, and the helical winding angle returns to >30°. In the short octameric duplex the unwinding caused by the platinum adduct opens both major and minor grooves, and the lower melting temperature of this duplex allows for more terminal base mobility. These factors facilitate the attack by a terminal guanine base from either the same or a second duplex on the G*pG* leading to the observed ICL.

In recent work [93] the effects of temperature and pH on duplexes containing a Pt([^{15}N] en)-*intra*strand cross-link were explored, revealing that there may be two distinct types of distorted helices formed. The platinated 14-mer duplex, d(ATACATG*G*TACATA)-(TATGTACCATGTAT), appeared to exist in an equilibrium between kinked and distorted helix forms. The two forms were detected by using a combination of [^{1}H, ^{15}N] and ^{1}H NMR and CD spectroscopy. The so-called kinked form was identified as a bent duplex containing the single platinum-1,2-*intra*strand adduct and having a structure similar to those previously reported [47,48]. This kinked duplex is the predominate species formed on platination of this oligomer [58]. The NMR spectral features of the distorted helix resemble those of platinated single-stranded DNA and appear under conditions of low pH (~4.5), low ionic strength (<10 mM NaCl) or high temperature (>310 K). This distorted form can be converted to the kinked duplex by raising the pH to 7. At higher ionic strength the kinked duplex predominates. It is inter-

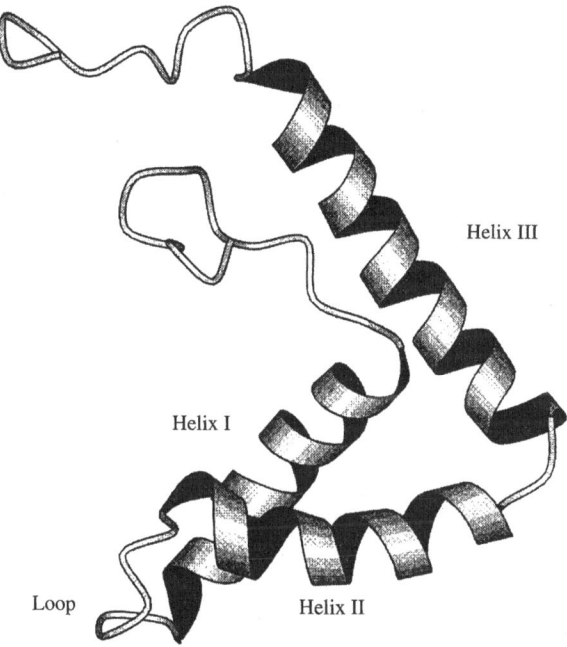

Fig. 12. MOLSCRIPT drawing of the solution structure of the DNA binding domain A of HMG 1. Redrawn from coordinates in [94]

esting that the distorted form, while resembling ssDNA, has some ordered structure. Its duplex character is confirmed by the detection of most of the imino resonances expected for the Watson-Crick base pairs. The structural transitions observed at low pH are reportedly due primarily to the protonation of cytosine residues [16,93], which could afford unwound duplexes or distorted (B-Z) helix transitions. Similar structural transitions were detected by using [^{1}H, ^{15}N] NMR spectroscopy during the binding of domain A of HMG1 to the kinked duplex [93].

Recently, the solution structures of both domain A (HMG1-A) [94] and domain B (HMG1-B) [95,96] of the DNA-binding protein HMG1 were determined by NMR spectroscopy. The two domains have a very similar global fold and structure. A MOLSCRIPT representation of HMG1-A is shown in Fig. 12. These domains bind specifically to cisplatin-modified DNA [97] (see Sect. 5). By monitoring changes in the known chemical shift values for individual residues of HMG1-A [94], insight was obtained into the possible mode of binding by these proteins to a platinum-modified duplex. The NMR spectrum of the 1:1 protein/DNA complex indicated the DNA to be in an intermediate exchange rate (on the NMR time scale) between free and protein-bound forms (K_d ~10^{-3}M), and that the duplex was distorted from the kinked form previously described [93]. The chemical shift changes indicated that the platinated duplex binds to the concave face of the domain, making contacts at the helix I, loop, helix II region. Such interactions would be consistent with protein-DNA contacts reported for the solution structures of two other HMG-domain proteins, LEF-1 [98] and SRY [99], bound to their native DNA recognition sequences. The interaction of HMG domains and other cellular proteins with cisplatin-modified DNA has been under intense study for several years [35,100] and may prove to be an important aspect of the mechanism of cytotoxicity of cisplatin in vivo. The following section describes recent investigations into this role of HMG-domain proteins, their recognition of, and binding to, cisplatin-modified DNA.

5
Interactions of Platinated DNA with Cellular Proteins

5.1
Discovery of Structure-Specific Recognition Protein SSRP1 and Its Interaction with Cisplatin-Modified DNA

The differential processing of cisplatin-DNA adducts and those of its inactive isomer trans-DDP suggests that the structural perturbations caused by platination may be recognized in a structure-specific way by cellular proteins. A search for mammalian proteins in cell extracts that bind cisplatin-modified DNA was therefore undertaken using a number of approaches with ^{32}P-radiolabeled, platinated DNA probes [101]. Gel mobility shift assays, in which native polyacrylamide gel electrophoresis (PAGE) was used to distinguish between free and protein-bound DNA on the basis of the retarded migration of the protein-DNA complex in the gel, were employed. A related method used in parallel work was the Southwestern blot analysis. This method uses SDS-PAGE to separate cellular proteins by molecular weight which are then electrophoretically transferred to nitrocellulose. The filters are subsequently probed with the radiolabeled cisplatin-modified DNA. The Southwestern method revealed that several proteins from human cell extracts with molecular weights of ~100 and ~28 kDa specifically recognized platinated DNA [34]. In the same study two cDNA clones were isolated from a human B-cell cDNA library which encoded portions of a protein that bound selectively to cisplatin-modified DNA. These partial length clones were used to screen additional libraries and ultimately provided a means of isolating and characterizing a full-length human cDNA clone which encoded for a structure-specific recognition protein (SSRP) termed SSRP1 [102]. This protein had the predicted [34] molecular mass of 81 kDa and contained a number of highly charged domains, including the 75 amino acid DNA-binding region known as the high-mobility group (HMG) domain [102]. Although these domains had been shown in other HMG-domain proteins to bind specific DNA sequences [103–105], this work demonstrated that such a domain could also recognize specifically DNA having altered structures.

Since the prototypical HMG domain occurs in the high-mobility group protein HMG1, it seemed likely that this protein would also bind cisplatin-modified DNA [36]. Through the combined use of Western and Southwestern blotting, the previously reported 28 kDa Pt-DNA recognition protein [34] was identified as a mixture of HMG1 and HMG2. This conclusion was reached independently [106]. DNA duplexes containing either a single site-specific cis-$[Pt(NH_3)_2\{d(GpG)$-$N7(1),N7(2)\}]$ or a cis-$[Pt(NH_3)_2\{d(GpTpG),N7(1),N7(3)\}]$ intrastrand cross-link were used to investigate the nature of the HMG1-platinated DNA interaction [36]. The specificity for the 1,2- over the 1,3-adduct was identical to that observed by Southwestern assays in cell extracts, the 1,2-adduct reacting specifically with HMG1 and showing a distinct gel mobility shift while the 1,3-adduct did not. HMG1 binds tightly to DNA containing the 1,2-intrastrand cross-link, with a K_d of 4×10^{-7}M vs. a K_d of 3×10^{-5}M for non-specific DNA binding.

The recognition of the major cisplatin-DNA adduct by HMG1 implied that the structural distortions, including bending and unwinding of the helix, caused by this platinum lesion are responsible for the protein-binding specificity. The degree of unwinding, more than helix bending, may be the key factor in protein recognition since gel electrophoresis measurements indicated that the 1,2- and 1,3-intrastrand adducts bend the DNA to a similar extent but have different effects on helix unwinding [18,19]. The recent high-resolution structures of both the 1,2- [48,70] and 1,3-adducts [83], however, reveal bend angle and other structural features which seem to differentiate these structures more than helical unwinding.

Not only do HMG-domain proteins recognize and bind pre-bent DNA, but they can further bend and distort the helix on binding. This observation was demonstrated through the use of a ring-closure assay in which short DNA fragments (87, 92, 100, and 123 bp) cyclized in the presence of HMG1 or HMG1-B and T4 DNA ligase [107]. The degree of cyclization was determined by gel electrophoresis, and fragments exposed to ligase in the absence of HMG1 showed no ring closure. This study indicated that the role of HMG1 may be one of structural distortion, bending DNA for nucleosome packaging, or of transcription mediation, facilitating binding of transcription factors to pre-bent DNA. Very recently a role was established for SSRP1 in transcriptional regulation of the human ε-globin gene. It is probable that SSRP1 acts as an architectural cofactor which helps to coordinate the assembly of a multiprotein gene regulatory complex by recognizing and bending a specific DNA sequence [108].

5.2
HMG-Domain Proteins Bind Specifically to Cisplatin-Modified DNA

5.2.1
HMG-Domain Proteins Bind and Further Distort Cisplatin-Modified DNA

In a study using circularly permuted 92 bp linear DNAs containing a single site-specific cis-$[Pt(NH_3)_2\{d(GpG)$-$N7(1),N7(2)\}]$ adduct, the structural effects of the binding of HMG-domain proteins to platinated DNA were investigated [109]. The use of these types of probes allows a helix bend angle to be calculated for short (<300–400 bp) duplexes based on the gel mobility of the protein-DNA complex in a native PAGE analysis. HMG1 bends the platinated DNA helix by 86 ±2°. This large bend angle may be compared to the cisplatin-induced value of ~39° determined previously by PAGE. This bend angle determined for specific binding of HMG1 suggests that, for long DNA, the two binding domains of HMG1 may act cooperatively. Indirect evidence for such activity was obtained by investigating a single HMG domain from HMG1. HMG1-B interacts with the platinated 92 bp duplex, bending it by 65–74°, somewhat less than the full length protein. In addition, this study revealed the bend locus to be centered near the platinum lesion, suggesting that the pre-bent DNA leads to binding of HMG domains to specific structural motifs which then induce a further distortion of the duplex.

The HMG domain is now a common motif that occurs in a variety of DNA-recognition proteins [110,111]. The more abundant chromosomal proteins HMG1 and HMG2 bind DNA with little or no sequence specificity. A second class of these proteins, which includes the transcriptional regulators LEF-1 (the lymphoid enhancer-binding factor) and SRY (the mammalian sex-determining factor), are characterized by a single HMG-domain binding specific DNA sequences. Both classes of proteins recognize distorted DNA structures such as four-way junctions [111] and were therefore examined for their ability to bind to cisplatin-modified DNA.

Using the circular permuted 92 bp DNAs mentioned above, the induced bend angle and Pt-DNA binding ability of a series of HMG-domain proteins were determined [109]. The yeast protein Ixr1 [112] and the human mitochondrial transcription factor mtTFA [113] are full-length HMG-domain proteins that bind to cisplatin-modified DNA in a manner comparable to that observed for HMG1 [109]. These proteins bend DNA by 68±6° and 89±5°, respectively, values similar to that observed for HMG1. In addition, the bend angles induced by the isolated HMG domains from LEF-1 and mSRY were determined to be 72±6° and ~50°. These values were comparable to those observed for the isolated B-domain from HMG1 of 65–74°. The observation that an isolated domain from an HMG-domain protein is sufficient to bind and distort further cisplatin-damaged DNA, and the general ability for both families of proteins to recognize platinated DNA, led to a number of detailed studies investigating the interaction of individual HMG-domains with short duplex DNAs containing the cisplatin 1,2-*intra*strand adduct. These proteins included a mouse testis-specific HMG-domain protein (tsHMG) [114], the human sex-determining factor SRY (hSRY) [115], and both domains from HMG1 [116], HMG1-A and HMG1-B. Table 2 compiles the results of these studies.

Table 2. Properties of HMG-domain protein interactions with cisplatin-modified DNA[a]

HMG-domain protein	HMG domains	$K_{d(app)}$ (M) Platinated DNA	Specificity (ρ)[b]	Bend angle (°)[c]	Pt-DNA probe[d]	Ref
HMG1	2	$3.7\pm2.0\times10^{-7}$	100	86 ± 2	100 bp/92 bp	36
HMG1-A	1	$1.6\pm0.2\times10^{-9}$	1000	nd	AGGA 15 bp	116
		$517\pm60\times10^{-9}$	3		CGGC 15 bp	
HMG1-B	1	$48\pm9\times10^{-9}$	nd	$65-74\pm4$	AGGT 15 bp/92 bp	116
		$1300\pm190\times10^{-9}$			CGGC 15 bp/92 bp	
Ixr1	1	$2.5\pm0.1\times10^{-7}$	10	68 ± 6	92 bp	112
tsHMG	2	$24\pm5\times10^{-9}$	230	nd	20 bp	114
tsHMG-A	1	$300\pm50\times10^{-9}$	20	nd	20 bp	114
mSRY	1	$\sim10^{-6}$	nd	~50	92 bp	109
hSRY	1	$120\pm10\times10^{-9}$	20	nd	20 bp	115
hSRY HMG domain	1	$4\pm0.7\times10^{-9}$	5	nd	20 bp	115
LEF-1	1	$\sim10^{-7}$	nd	72 ± 6	92 bp	109
LEF-1/DNA	1	$\sim10^{-9}$		117	15 bp	98
hSRY /DNA	1	$15\pm3\times10^{-9}$	10-20	70-80	20 bp/8 bp	99

a The LEF-1/DNA and hSRY/DNA parameters for proteins bound to their native recognition sequences are included for comparison. The K_d value is for a duplex DNA containing the particular sequence recognized by the protein; the bend angle is from the NMR structure of each complex as discussed in the text.

b The specificity number refers to the ratio K_d(unplatinated DNA)/K_d(cisplatin-modified DNA).

c As determined using the circular permuted DNA assay in [109].

d The first number is the DNA length used for K_d determination, the second is for bend angle determination.

nd: not determined

5.2.2
The Structures of Two HMG-Domain Proteins Bound to Their Native DNA Sequences

A high-resolution structure of cisplatin-modified DNA bound to a cellular protein is not yet available. Nevertheless, some insight can be provided into the pertinent intermolecular interactions through examination of the solution structures of HMG1-A [94] and HMG1-B [95,96] and of two different sequence-specific HMG-domain proteins LEF-1 [98] and hSRY [99] bound to their native DNA recognition sequences. The individual DNA-binding regions of HMG1 recognize, bind with similar affinities ($K_d\sim10^{-9}$), and further distort cisplatin-modified DNA of various lengths. Solution structures of these two domains were recently determined independently to have very similar secondary structures consisting of three α-helices arranged in a distorted L-shape providing a concave surface for DNA binding [94,96]. The structure of HMG1-A is depicted in

Fig. 12. Furthermore, both the C- and N-termini are at the ends of disordered loop structures which may be free to create additional intermolecular contacts. The acidic C-terminus of the HMG-domain protein HMG-D, which is isostructural to HMG1-A, is indeed important for structure recognition [117].

The structure of the isolated domain undergoes relatively minor structural perturbations upon binding to duplex DNA [98,99]. This result is clearly illustrated by the structure of LEF-1 bound to a 15-mer oligonucleotide d(CAC-**CCTTTGAAG**CTC) · d(GAG**CTTCAAAGG**GTG) containing the internal recognition sequence shown in boldface type [98]. The structure of this HMG-domain-DNA complex is shown in Fig. 13.

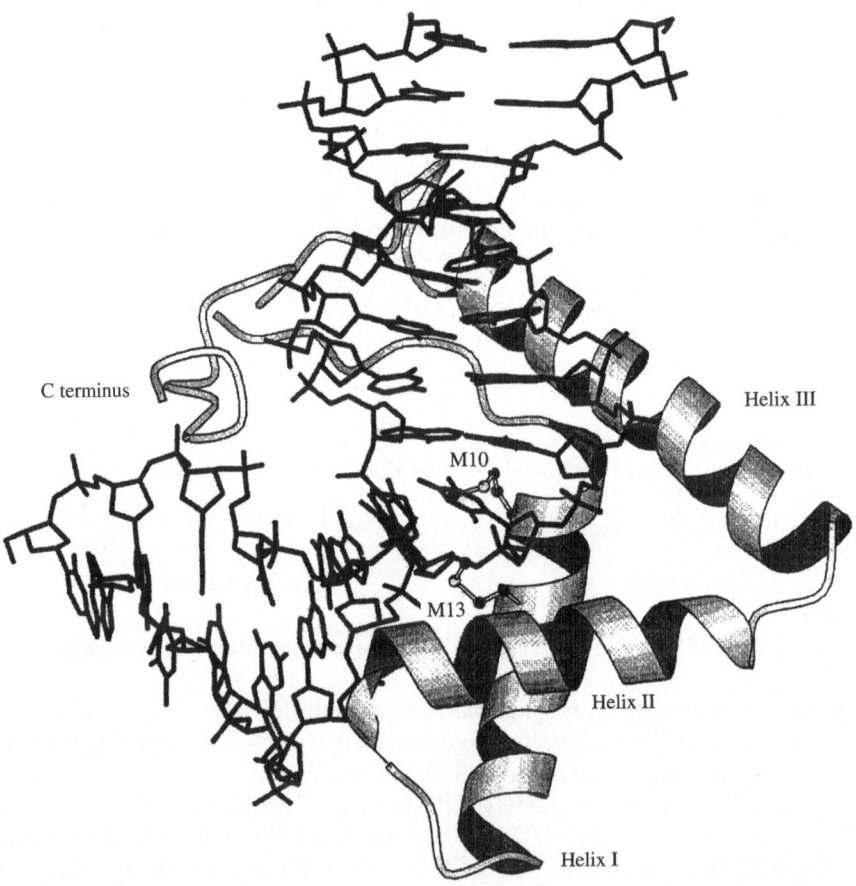

Fig. 13. MOLSCRIPT representation of the structure of the LEF-1 HMG-domain complex with a 15-mer duplex d(CACCCTTTGAAGCTC)·d(GAGCTTCAAAGGGTG). The two methionine side chains M10 and M13 which insert the A23–A24 base step are indicated. Redrawn from coordinates provided in [98]

The DNA in this complex binds to the concave hydrophobic surface of the LEF-1 domain primarily contacting helix I and helix II, and the C-terminus of helix III which binds across the major groove contributing to the structure recognition. The DNA forms a large number of contacts with the protein across its widened (~11 Å) and flattened (~1 Å) minor groove. The latter is reminiscent of that observed in the structures of the cisplatin-modified duplexes S1, S2, and S4. Binding of the domain to DNA involves severe bending and unwinding of the helix. The helix is bent towards the major groove with an overall bend angle of ~117°, primarily due to local base pair distortions over the region C5–G26 to T8–A23. The duplex is also unwound in this region, with average twist angles of 19–24°. The major recognition motif of this protein appears to involve extensive side-chain contacts with the DNA from the C4–G27 through the A11–T20 base pairs. Most of the DNA contacts involve the sugar-phosphate backbone, but two methionine residues M10 and M13 interact with the base stacking at the A23–A24 step. These residues are labeled in Fig. 13. The roll of 64° observed at this base step is similar to that induced at the cisplatin-d(G*pG*) site in solution [48,49]. The DNA structural parameters in this complex are so similar to those observed for the recent high-resolution structures of cisplatin-modified DNA (see Table 1) that it seems that the cisplatin adduct pre-bends DNA in such a way that the protein could recognize it without further modification [48]. Although earlier gel electrophoresis studies showed that binding of HMG-domain proteins can further bend platinated DNA [109] from the apparently more shallow 32–40° bend angle of the platinum-modified duplex alone [18], these more recent NMR studies reveal that the DNA may be sufficiently distorted by the cisplatin adduct. Structural studies on complexes of HMG-domain proteins with cisplatin-modified DNA should provide the insight required to address some of these issues.

5.2.3
Specific Interactions Between a Single HMG Domain and Cisplatin-Modified DNA

In an effort to identify specific contacts made between HMG-domain proteins and cisplatin-modified DNA, a series of cisplatin analogs containing ligands with photoreactive aryl azide groups were prepared for use in cross-linking studies [118]. The use of ligands which can be photoactivated allows DNA-protein complexes to form under typical conditions. Photogeneration of the highly reactive nitrene moiety would covalently cross-link protein residues in close proximity to the platinum center. During the course of this study, it was shown in control experiments that cisplatin alone undergoes a photo-cross-linking reaction with HMG1 upon irradiation with 300 nm light, and does so with a higher efficiency than the complexes containing aryl azide ligands. Since single HMG domains can bind to short duplexes containing a single cisplatin-GpG adduct, HMG1-B was photo-cross-linked to a platinated 15 bp duplex in this manner. The use of a discrete adduct allowed the nature of the Pt-DNA-protein cross-link to be investigated. Proteolysis of the resulting protein-DNA complex, followed by gel purification and N-terminal sequencing of the fragments bound to the DNA, indicated that a covalent adduct had formed between the platinum atom and Lys-6, an amino acid somewhat removed from the helical regions of the protein. Such residues near the flexible N-terminus of the protein can interact with various regions of the bound DNA, and examples of such interactions with both N- and C-termini have been reported for other HMG-domain DNA complexes [98]. These results provided the first direct structural information about Pt-DNA-HMG domain interactions. Whereas the primary protein-DNA contacts may occur along the helices of the domains [119], the flexible ends of the peptide chain might form contacts that facilitate specific recognition of the cisplatin 1,2-*intra*strand lesion.

The diversity of HMG-domain proteins, all of which bind platinated DNA, suggests that the interaction is structure, rather than sequence, specific. A recent study has shown, however, that there is a sequence-specific component of recognition for the two DNA-binding domains of HMG1 [116]. A series of 15 bp oligonucleotides, each containing a single cisplatin-1,2-*intra*strand adduct, were prepared having the general sequence d(CCTCTCN$_1$G*G*N$_2$TCTTC) · (GAAG-AN$_3$CCN$_4$GAGAGG) and where N$_1$ and N$_2$ were combinations of A, C, or T. Nine probes were prepared and used to investigate HMG1-A and HMG1-B binding to cisplatin-modified DNA. HMG1-A bound much more tightly than HMG1-B to any of the platinated DNAs, having typically a 10- to100-fold lower K_d value. Both proteins bound preferentially to DNA having A/T flanking sequences, but the base pair preference was not the same for each protein. Whereas HMG1-A binds the AG*G*A sequence most strongly with K_d=1.6 nM, the AG*G*T internal sequence is favored by HMG1-B (K_d=50 nM). This result is consistent with the observation that the HMG-domain proteins can further bend and distort platinated DNA [18,19,109]. The more flexible A-T base pairs flanking the platinum lesion would facilitate such bending of the DNA helix.

The tighter binding of HMG1-A is also accompanied by a much more distinct flanking sequence specificity. The 3'-flanking nucleotide determined the sequence preference, with the binding affinity increasing from dC to T to dA. The second binding domain HMG1-B exhibited a slightly different specificity, where the affinity decreased with N$_2$=T>dA=dC. The different binding abilities of the two domains in a full-length protein suggests a mechanism for sequence recognition. The less discriminating, weaker binding, domain B may recognize general shape distortions caused by cisplatin adducts and the more strongly binding domain A may select more easily distorted sequences. Binding of other proteins to cisplatin-modified DNA may also exhibit a sequence dependence which would allow particular lesions to persist if repair shielding, discussed next, were to occur.

5.2.4
HMG-Domain Proteins Block Excision Repair of Cisplatin-DNA Adducts Both In Vitro and In Vivo

The interaction of cisplatin-modified DNA with other cellular proteins is a burgeoning subject that has recently been reviewed in detail [100,120]. One consequence of such binding, just proposed for HMG-domain proteins, is inhibition of nucleotide excision repair [121]. Cells can become resistant to cisplatin by enhanced repair of platinum-DNA lesions. The role proposed for HMG-domain protein binding to cisplatin-modified DNA would be to block repair of the 1,2-*intra*strand cross-links. In such a mechanism, binding of HMG domains specifically to the major cisplatin adducts on DNA inhibits recognition of and repair by a cellular repair apparatus such as that used to perform nucleotide excision repair (NER). The NER system in crude HeLa human cell extracts repairs both the 1,2-d(GpG) and the 1,3-d(GpTpG) *intra*strand cross-links, with the latter being repaired more efficiently [121].

HMG-domain proteins bind to both the 1,2-d(GpG) or 1,2-d(ApG) *intra*strand adducts, but not to the 1,3-d(GpTpG) adduct. A study using purified repair components was undertaken to investigate the effect of these HMG-domain proteins on excision repair [122]. Both HMG1 and tsHMG, as well as the isolated HMG domain HMG1-B, specifically inhibited repair of either the 1,2-d(GpG) or 1,2-d(ApG) lesion on a 156 bp probe. Recently, similar work showed that hSRY can also block nucleotide excision repair [115]. Interestingly, the purified excinuclease of the reconstituted NER system also repaired the 1,3-d(GpTpG) adduct more efficiently than either of the 1,2-adducts, indicating that the structure might influence the repair activity of the excinuclease system.

The results of these experiments show that the basal level of HMG1-domain proteins present in HeLa cell extracts is insufficient to block repair of the major cisplatin adduct on DNA, but that elevated levels could specifically inhibit repair of the 1,2-*intra*strand lesion, providing support for the repair-shielding model of the role of HMG-domain proteins in mediating the cytotoxicity of cisplatin. Studies in vivo revealed that yeast cells, in which the HMG-domain protein Ixr1 was deleted, were less sensitive to cisplatin than the wild-type cells [123]. The Ixr1 protein binds strongly to cisplatin-modified DNA, like many other HMG-domain proteins [112,123], and a detailed study of yeast lacking Ixr1 indicated that its HMG domain may block the excision repair of the cisplatin adducts in vivo [124]. Studies in which HMG-domain proteins are over-expressed in cells, to mimic elevated production levels, will further address the repair-shielding hypothesis.

6
Summary and Future Outlook

The quest to understand the molecular basis for the anticancer activity of *cis*-DDP has spanned the thirty years since its discovery. The therapeutic limitations of cisplatin have driven the development of platinum-based derivatives ranging from minor leaving group modifications of the parent compound to the development of active trans and dinuclear complexes. The latter were developed, as were many other compounds, in an effort to circumvent both intrinsic and acquired cisplatin-resistance in tumor cells. One mechanism of resistance is the enhanced repair of platinum adducts. The recognition of platinum-DNA structural deformations by cellular proteins has been demonstrated both to enhance and to inhibit cisplatin efficacy. Nucleotide excision repair can remove the major cisplatin adduct from DNA. DNA-binding proteins known as HMG-domain proteins recognize the cisplatin-1,2-*intra*strand DNA cross-link specifically and inhibit NER. The binding of these proteins, as well as individual HMG domains, to DNA depends on the specific structural perturbation caused by 1,2-*intra*strand adducts. High-resolution structure determinations of a number of both major and minor cisplatin-DNA adducts have revealed details of the cisplatin lesion responsible for cellular protein recognition. These protein-DNA interactions will be complemented in the future by a high-resolution structure determination of a discrete HMG domain bound to cisplatin-modified DNA. Molecular biological studies of the effects on cisplatin sensitivity by in vivo expression of HMG-domain proteins are awaited with interest.

Acknowledgements. This work was supported by a National Cancer Institute Grant CA34992. A.G. is a recipient of a National Institutes of Health postdoctoral fellowship.

References

1. Pil P, Lippard SJ (1997) Cisplatin and related drugs. In: Bertino JR (ed) Encyclopedia of cancer. Academic Press, San Diego, p 392
2. Loehrer PJ, Einhorn LH (1984) Ann Inter Med 100:704
3. Pinedo HM, Schornagel JH (eds) (1996) Platinum and other metal coordination compounds in cancer chemotherapy. Plenum Press, New York
4. Chu G (1994) J Biol Chem 269:787
5. Rosenberg B, Van Camp L, Krigas T (1965) Nature 205:698
6. Rosenberg B, Van Camp L, Trosko JE, Mansour VH (1969) Nature 222:385
7. van der Vijgh WJF (1991) Clinical Pharmacokinetics 21:242
8. Kelland LR, Abel G, McKeage MJ, Jones M, Goddard PM, Valenti M, Morrer BA, Harrap KR (1993) Cancer Research 53:2581
9. Hartwig JF, Lippard SJ (1992) J Am Chem Soc 114:5646
10. Fichtinger-Schepman AMJ, van der Veer JL, den Hartog JHJ, Lohman PHM, Reedijk J (1985) Biochemistry 24:707
11. Fichtinger-Schepman AMJ, Dijt FT, de Jong WH, van Oosterom AT, Berends F (1988) In vivo *cis*-diamminedichloroplatinum(II)-DNA adduct formation and removal as measured with immunochemical techniques. In: Nicolini M (ed) Platinum and other metal coordination compounds in cancer chemotherapy. Nijhoff, Boston, p 33
12. Sherman SE, Gibson D, Wang AH-J, Lippard SJ (1985) Science 230:412
13. den Hartog JHJ, Altona C, Chottard J-C, Girault J-P, Lallemand J-Y, de Leeuw FAAM, Marcelis ATM, Reedijk J (1982) Nucleic Acids Res 10:4715
14. Admiraal G, van der Veer JL, de Graaff RAG, den Hartog JHJ, Reedijk J (1987) J Am Chem Soc 109:592

15. den Hartog JHJ, Altona C, van Boom JH, van der Marel GA, Haasnoot CAG, Reedijk J (1985) J Biomol Struct Dynam 2:1137
16. Herman F, Kozelka J, Stoven V, Guittet E, Girault J-P, Huynh-Dinh T, Igolen J, Lallemand J-Y, Chottard J-C (1990) Eur J Biochem 194:119
17. Rice JA, Crothers DM, Pinto AL, Lippard SJ (1988) Proc Natl Acad Sci USA 85:4158
18. Bellon SF, Lippard SJ (1990) Biophys Chem 35:179
19. Bellon SF, Coleman JH, Lippard SJ (1991) Biochemistry 30:8026
20. Sanderson BJS, Ferguson LR, Denny WA (1996) Mutat Res-Fundam Mol Mech Mutagen 355:59
21. Gately DP, Howell SB (1993) Brit J Cancer 67:1171
22. Matsumoto T, Endoh K, Akamatsu K, Kamisango K, Mitsui H, Koizumi K, Morikawa K, Koizumi M, Matsuno T (1991) Brit J Cancer 64:41
23. Holford J, Raynaud F, Murrer BA, Grimaldi K, Hartley JA, Abrams M, Kelland LR (1998) Anti-Cancer Drug Des 13:1
24. Holford J, Sharp SY, Murrer BA, Abrams M, Kelland LR (1998) Brit J Cancer 77:366
25. Kelland LR (1993) Crit Rev Oncol-Hematol 15:191
26. Bleiberg H (1998) Brit J Cancer 77(S4):1
27. Cvitkovic E (1998) Brit J Cancer 77(S4):8
28. McKeage MJ, Abel G, Kelland LR, Harrap KP (1994) Brit J Cancer 69:1
29. Mellish KJ, Kelland LR (1994) Cancer Res 54:6194
30. Novakova O, Vrana O, Kiseleva VI, Brabec V (1995) Eur J Biochem 228:616
31. McKeage MJ, Mistry P, Ward J, Boxall FE, Loh S, O'Neill C, Ellis P, Kelland LR, Morgan SE, Murrer BA, Santabarbara P, Harrap KR, Judson IR (1995) Cancer Chemother Pharmacol 36:451
32. Harrap KR, Murrer BA, Giandomenico C, Morgan SE, Kelland LR, Jones M, Goddard PM, Schurig J (1991) Ammine/amine platinum(IV) dicarboxylates: a novel class of complexes which circumvent intrinsic cisplatin resistance. In: Howell SB (ed) Platinum and other metal coordination complexes in cancer chemotherapy. Plenum Press, New York, p 391
33. Sandman KS, Fuhrmann P, Lippard SJ (1998) J Biol Inorg Chem 3:74
34. Toney JH, Donahue BA, Kellett PJ, Bruhn SL, Essigmann JM, Lippard SJ (1989) Proc Natl Acad Sci USA 86:8328
35. Whitehead JP, Lippard SJ (1996) Proteins that bind to and mediate the biological activity of platinum anticancer drug-DNA adducts. In: Sigel A, Sigel H (eds) Metal ions in biological systems. Marcel Dekker Inc, New York, p 687
36. Pil PM, Lippard SJ (1992) Science 256:234
37. Sandman KE, Lippard SJ (1999) Methods for sreening the potential antitumor activity of platinum compounds in combinatorial libraries. In: Lippert B (ed) Thirty years of cisplatin: chemistry and biochemistry of a leading anticancer drug (in press)
38. Kelland LR, Barnard CFJ, Mellish KJ, Jones M, Goddard PM, Valenti M, Bryant A, Murrer BA, Harrap KR (1994) Cancer Res 54:5618
39. Kelland LR, Barnard CFJ, Evans IG, Murrer BA, Theobald BRC, Wyer SB, Goddard PM, Jones M, Valenti M, Bryant A, Rogers PM, Harrap KR (1995) J Med Chem 38:3016
40. Kraker AJ, Hoeschele JD, Elliott WL, Showalter HDH, Sercel AD, Farrell NP (1992) J Med Chem 35:4526
41. Farrell N (1993) Cancer Invest 11:578
42. Zou Y, van Houten B, Farrell N (1994) Biochemistry 33:5404
43. Kasparkova J, Mellish KJ, Qu Y, Brabec V, Farrell N (1996) Biochemistry 35:16705
44. Zaludova R, Zakovska A, Kasparkova J, Balcarova Z, Kleinwachter V, Vrana O, Farrell N, Brabec V (1997) Eur J Biochem 246:508
45. Sherman SE, Gibson D, Wang AH-J, Lippard SJ (1988) J Am Chem Soc 110:7368
46. Takahara PM, Rosenzweig AC, Frederick CA, Lippard SJ (1995) Nature 377:649
47. Yang D, van Boom SSGE, Reedijk J, van Boom JH, Wang AH-J (1995) Biochemistry 34:12912

48. Gelasco A, Lippard SJ (1998) Biochemistry 37:9230
49. Dunham SU, Dunham SU, Turner CJ, Lippard SJ (1998) J Am Chem Soc 120:5395
50. Johnson NP, Hoeschele JD, Rahn RO (1980) Chem-Biol Interactions 30:151
51. Bancroft DP, Lepre CA, Lippard SJ (1990) J Am Chem Soc 112:6860
52. Orton DM, Gretton VA, Green M (1993) Inorg Chim Acta 204:265
53. Fichtinger-Schepman AMJ, Lohman PHM, Reedijk J (1982) Nucleic Acids Res 10:5345
54. Elmroth SKC, Lippard SJ (1994) J Am Chem Soc 116:3633
55. Hambley TW (1997) Coord Chem Rev 166:181
56. Bloemink MJ, Reedijk J (1996) Cisplatin and derived anticancer drugs: mechanism and current status of DNA binding. In: Sigel A, Sigel H (eds) Metal ions in biological systems. Marcel Dekker Inc, New York, p 641
57. Barnham KJ, Berners-Price SJ, Frenkiel TA, Frey U, Sadler PJ (1995) Angew Chem Int Ed Engl 34:1874
58. Berners-Price SJ, Barnham KJ, Frey U, Sadler PJ (1996) Chem Eur J 2:1283
59. Dewan J (1984) J Am Chem Soc 106:7239
60. Hongo A, Seki S, Akiyama K, Kudo T (1994) Int J Biochem 26:1009
61. Cohen GL, Bauer WR, Barton JK, Lippard SJ (1979) Science 203:1014
62. Blommaert FA, van Dijk-Knijnenburg HCM, Dijt FJ, den Engelse L, Baan RA, Berends F, Fichtinger-Schepman AMJ (1995) Biochemistry 34:8474
63. Fichtinger-Schepman AMJ, van Dijk-Knijnenburg HCM, van der Velde-Visser SD, Berends F, Baan RA (1995) Carcinogenesis 16:2447
64. Salazar I, Tarrago-Litvak L, Gil L, Litvak S (1982) FEBS Lett 138:45
65. Olivero OA, Semino C, Kassim A, Lopez-Larraza DM, Poirier MC (1995) Mutat Res Lett 346:221
66. Olivero OA, Chang PK, Lopez-Larraza DM, Semino-Mora MC, Poirier MC (1997) Mutat Res Genet Toxicol Environ Mutagen 391:79
67. Counter CM (1996) Mutat Res Reviews in Genetic Toxicology 366:45
68. Counter CM, Hirte HW, Bacchetti S, Harley CB (1994) Proc Natl Acad Sci USA 91:2900
69. Ishibashi T, Lippard SJ (1998) Proc Natl Acad Sci USA 95:4219
70. Takahara PM, Frederick CA, Lippard SJ (1996) J Am Chem Soc 118:12309
71. Huang H, Zhu L, Reid BR, Drobny GP, Hopkins PB (1995) Science 270:1842
72. Paquet F, Perez C, Leng M, Lancelot G, Malinge J-M (1996) J Biomol Struct Dyn 14:67
73. Kraulis PJ (1991) J Appl Cryst 24:946
74. Sundquist WI, Lippard SJ, Stollar BD (1986) Biochemistry 25:1520
75. van Boom SSGE, Yang D, Reedijk J, van der Marel GA, Wang AH-J (1996) J Biomol Struct Dyn 13:989
76. Dunham SU, Lippard SJ (1995) J Am Chem Soc 117:10702
77. Lemaire M-A, Schwartz A, Rahmouni AR, Leng M (1991) Proc Natl Acad Sci USA 88:1982
78. Sip M, Schwartz A, Vovelle F, Ptak M, Leng M (1992) Biochemistry 31:2508
79. Malinge J-M, Pérez C, Leng M (1994) Nucleic Acids Res 22:3834
80. Wang AH-J, Quigley GJ, Kolpak FJ, Crawford JL, van Boom JH, van der Marel G, Rich A (1979) Nature 282:680
81. Rappé AK, Casewit CJ (1997) Molecular mechanics across chemistry, 1st edn. University Science Books, Sausalito, CA
82. den Hartog JHJ, Altona C, van der Elst H, van der Marel GA, Reedijk J (1985) Inorg Chem 24:983
83. van Garderen CJ, van Houte LPA (1994) Eur J Biochem 225:1169
84. Lepre CA, Chassot L, Costello CE, Lippard SJ (1990) Biochemistry 29:811
85. Yang D, van Boom SSGE, Reedijk J, van Boom JH, Farrell N, Wang AH-J (1995) Nature Struct Biol 2:577
86. Yang D, Wang AH-J (1996) Prog Biophys Molec Biol 66:81
87. Ferrari S, Harley VR, Pontiggia A, Goodfellow PN, Lovell-Badge R, Bianchi ME (1992) EMBO J 11:4497

88. Comess KM, Costello CE, Lippard SJ (1990) Biochemistry 29:2102
89. Dalbiès R, Payet D, Leng M (1994) Proc Natl Acad Sci USA 91:8147
90. Dalbiès R, Boudvillain M, Leng M (1995) Nucleic Acids Res 23:949
91. Pérez C, Leng M, Malinge J-M (1997) Nucleic Acids Res 25:896
92. Gelasco A, Dunham SU, Lippard SJ unpublished results
93. Berners-Price SJ, Corazza A, Guo Z, Barnham KJ, Sadler PJ, Ohyama Y, Leng M, Locker D (1997) Eur J Biochem 243:782
94. Hardman CH, Broadhurst RW, Raine ARC, Grasser KD, Thomas JO, Laue ED (1995) Biochemistry 34:16596
95. Read CM, Cary PD, Crane-Robinson C, Driscoll PC, Norman DG (1993) Nucleic Acids Res 21:3427
96. Weir HM, Kraulis PJ, Hill CS, Raine ARC, Laue ED, Thomas JO (1993) EMBO J 12:1311
97. Chow CS, Barnes CM, Lippard SJ (1995) Biochemistry 34:2956
98. Love JJ, Li X, Case DA, Giese K, Grosschedl R, Wright PE (1995) Nature 376:791
99. Werner MH, Huth JR, Gronenborn AM, Clore GM (1995) Cell 81:705
100. Zamble DB, Lippard SJ (1999) The response of Cellular proteins to Cisplatin-damaged DNA. In: Lippert B (ed) Thirty tears of cisplatin: chemistry and biochemistry of a leading anticancer drug (in press)
101. Bruhn SL, Toney JH, Lippard SJ (1990) Prog Inorg Chem 38:477
102. Bruhn SL, Pil PM, Essigmann JM, Housman DE, Lippard SJ (1992) Proc Natl Acad Sci USA 89:2307
103. Sinclair AH, Berta P, Palmer MS, Hawkins JR, Griffiths BL, Smith MJ, Foster JW, Frischauf A-M, Lovell-Badge R, Goodfellow PN (1990) Nature 346:240
104. Travis A, Amsterdam A, Belanger C, Grosschedl R (1991) Genes Dev 5:880
105. Parisi MA, Clayton DA (1991) Science 252:965
106. Hughes EN, Engelsberg BN, Billings PC (1992) J Biol Chem 267:13520
107. Pil PM, Chow CS, Lippard SJ (1993) Proc Natl Acad Sci U S A 90:9465
108. Dyer MA, Hayes PJ, Baron MH (1998) Mol Cell Bio 18:2617
109. Chow CS, Whitehead JP, Lippard SJ (1994) Biochemistry 33:15124
110. Bianchi ME, Beltrame M, Falciola L (1992) In: Eckstein F, Lilley DMJ (eds) Nucleic acids and molecular biology. Springer-Verlag, Berlin Heidelberg, p 112
111. Grosschedl R, Giese K, Pagel J (1994) Trends Genet 10:94
112. McA'Nulty MM, Whitehead JP, Lippard SJ (1996) Biochemistry 35:6089
113. Fisher RP, Lisowsky T, Parisi MA, Clayton DA (1992) J Biol Chem 267:3358
114. Ohndorf U-M, Whitehead JP, Raju NL, Lippard SJ (1997) Biochemistry 36:14807
115. Trimmer EE, Zamble DB, Lippard SJ, Essigmann JM (1998) Biochemistry 37:352
116. Dunham SU, Lippard SJ (1997) Biochemistry 35:11428
117. Payet D, Travers A (1997) J Mol Biol 266:66
118. Kane SA, Lippard SJ (1996) Biochemistry 35:2180
119. Read CM, Cary PD, Preston NS, Lnenicek-Allen M, Crane-Robinson C (1994) EMBO J 13:5639
120. Zamble DB, Lippard SJ (1995) Trends Biochem Sci 20:435
121. Huang J-C, Zamble DB, Reardon JT, Lippard SJ, Sancar A (1994) Proc Natl Acad Sci USA 91:10394
122. Zamble DB, Mu D, Reardon JT, Sancar A, Lippard SJ (1996) Biochemistry 35:10004
123. Brown SJ, Kellett PJ, Lippard SJ (1993) Science 261:603
124. McA'Nulty MM, Lippard SJ (1996) Mutat Res-DNA Repair 362:75

Development of an Orally Active Platinum Anticancer Drug: JM216

C.F.J. Barnard[1], F.I. Raynaud[2], L.R. Kelland[2*]

[1] Johnson Matthey Technology Centre, Blount's Court, Sonning Common, Reading RG4 9NH, UK
[2] CRC Centre for Cancer Therapeutics, The Institute of Cancer Research, 15 Cotswold Road, Sutton, Surrey SM2 5NG, UK
E-mail: [1] barnacfj@matthey.com, [2] floren@icr.ac.uk, [2*] lloyd@icr.ac.uk

Following the successful introduction of the less toxic analogue of cisplatin, carboplatin, into clinical practice in the early 1980s, a collaborative programme of research was established between Johnson Matthey, the Institute of Cancer Research and Bristol Myers Squibb. Its aim was to discover and develop an orally active platinum drug possessing at least comparable antitumour activity to that of cisplatin but a toxicological profile reminiscent of carboplatin. A new class of platinum compounds synthesized specifically to circumvent the poor gastrointestinal absorption of cisplatin and carboplatin possessed the properties of having a relatively low molecular weight, were lipophilic, neutral, kinetically inert and acid stable. The resulting lead compound JM216 (bis-acetato-ammine dichlorocyclohexylamine platinum IV), an example of the Pt(IV) mixed ammine/amine dicarboxylate dichloride series, entered clinical trials at the Royal Marsden Hospital, London, in 1992. Preclinically, JM216 was demonstrated to possess oral antitumour activity in mice bearing the ADJ/PC6 plasmacytoma and a panel of human ovarian carcinoma xenografts broadly equivalent to that observed for intravenously administered cisplatin or carboplatin. Oral antitumour activity was greater when using a daily×5 split dose schedule versus weekly dosing, probably as a result of saturable oral absorption. In vitro against panels of cisplatin-sensitive and -resistant human tumour cell lines, JM216 exhibited a similar potency to that of cisplatin and was able to circumvent acquired cisplatin resistance due to reduced drug transport. The drug's toxicological profile in rodents was similar to that of carboplatin with myelosuppression being dose-limiting with no obvious nephro- or neurotoxicity. The metabolism of JM216 is complex with up to six metabolites being formed; the major metabolite in man being cis-ammine dichloro(cyclohexylamine) platinum (II) (JM118). Glutathione conjugation represents a major deactivation pathway for JM216. The initial single dose phase I clinical trial was incapable of defining a maximum tolerated dose (MTD) due to absorption-limited non-linear pharmacokinetics. A second phase I trial daily for 5 days, showed dose-limiting toxicities of thrombocytopenia and neutropenia. Recommended phase II doses were 100 and 120 mg/m^2/day×5 for previously treated and untreated patients, respectively. Phase II trials are currently ongoing in a number of tumour types including prostate, ovarian and lung.

Keywords. Platinum, Oral, Chemistry, Biology, Pharmacology

1
Introduction

1.1
Goals for the Development of Platinum Drugs

Following its first marketing approval in 1978, cisplatin has proved to be one of the most successful anti-cancer drugs yet developed. It is widely used in combination therapy for many tumours, although the major therapeutic benefits have been in testicular and ovarian cancer. However, even before it was marketed, clinical trials indicated that there were major difficulties with its use related to its toxicity. Cisplatin produces a wide range of side effects including kidney toxicity and neurotoxicity and induces severe nausea and vomiting. During the 1970s it was found that the kidney toxicity could be reduced by the use of pre- and posthydration of the patient during administration of the drug, and subsequently new anti-emetic agents ($5HT_3$-antagonists such as Ondansetron and Granisetron) have been developed to ameliorate nausea and vomiting. Nonetheless, it has been apparent since the launch of cisplatin that alternative platinum drugs could offer significant benefits over cisplatin.

Early studies of alternative platinum compounds concentrated on efforts to reduce the toxicity of the drug while retaining the therapeutic benefits, thus making its administration easier and potentially extending opportunities for the use of the drug in combination with other anti-cancer agents. The first of this second-generation of anti-cancer drugs to be approved was carboplatin, first marketed in 1986 [1]. The dose-limiting toxicity of this compound is haematological, and at the normal maximum tolerated dose there is essentially no kidney toxicity and very little neurotoxicity. (Subsequent developments in the use of stimulants for hematopoietic stem cells have allowed dose intensification for carboplatin, when non-haematological toxicities become more evident.)

Of the many platinum compounds that have been tested clinically as drugs with potentially reduced toxicity compared with cisplatin, a number have now received limited marketing approval (Fig. 1) – e.g. 254-S (nedaplatin) in Japan and oxaliplatin in France (see [2] for a review). However, to date, carboplatin is the only second-generation analogue in widespread use.

The clinical trials of carboplatin established that it shows a high degree of cross resistance with cisplatin [3]. While many tumours initially respond to cisplatin treatment, this therapy is often not completely successful such that the tumour re-grows and is then more resistant to the drugs used initially. An alternative compound which does not show cross-resistance with cisplatin has therefore become another major goal in the development of platinum drugs.

The structure-activity 'rules' for platinum drugs were first defined by Cleare and colleagues following testing of a wide range of platinum complexes [4]. However, in order to obtain compounds which might lack cross resistance with cisplatin much effort has been devoted to compounds which fall outside these guidelines and many are now receiving detailed evaluation. Some compounds of

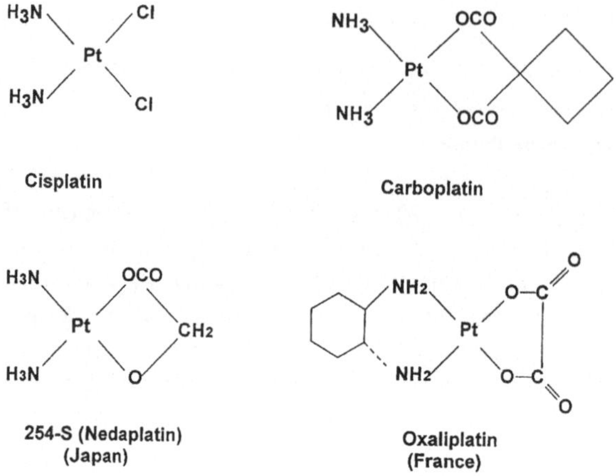

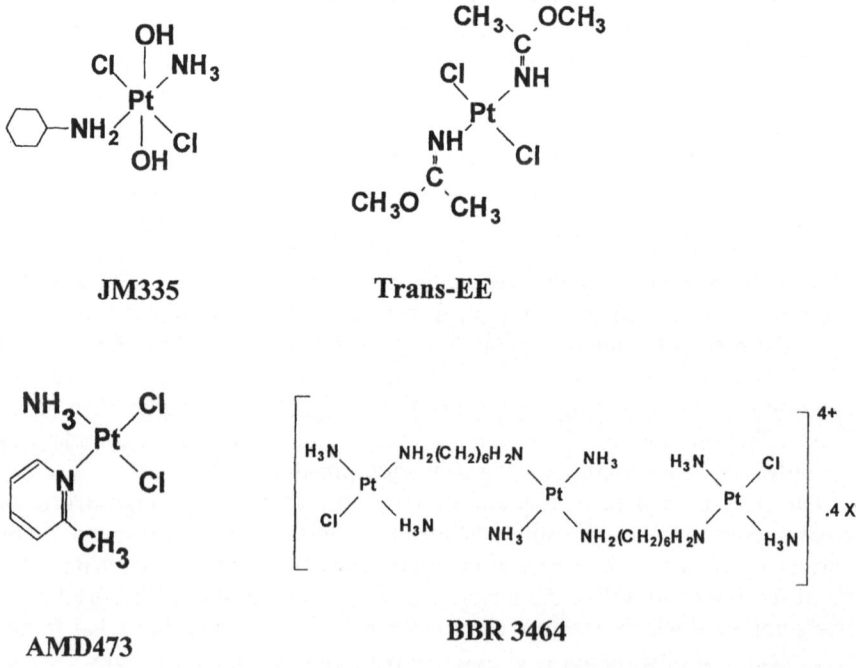

Cisplatin

Carboplatin

254-S (Nedaplatin)
(Japan)

Oxaliplatin
(France)

Fig. 1. Platinum drugs with marketing approval

JM335

Trans-EE

AMD473

BBR 3464

Fig. 2. Platinum compounds of interest in terms of preclinical activity/mechanism of action

interest are shown in Fig. 2. Some of these are *trans*-complexes (e.g. JM335) which flout the most firmly established of the structure-activity guidelines; i.e. that the activity of the *cis*-isomers greatly exceeds that of the trans forms. With the exception of AMD473 which began phase I clinical trials in 1997, none of these "third-generation" compounds has yet been evaluated clinically so it is presently unknown whether these interesting pre-clinical results will lead to the development of new drugs.

Cisplatin and all the second-generation platinum drugs are administered by intravenous infusion. The ability to deliver the drug orally would allow much greater flexibility in dosing and increase the potential for the use of platinum drugs, especially in palliative care. However, the physical properties of cisplatin rule out the possibility of an effective oral formulation. While antitumour activity for cisplatin following oral administration can be demonstrated (see below), the low level of absorption makes this impractical. Carboplatin and most of the other second-generation compounds have greater water solubility than cisplatin (desirable for an intravenously administered drug of lower potency than cisplatin) and very low organic/aqueous partition coefficients which would be expected to lead to low absorption. This was confirmed by a brief clinical study of carboplatin given p.o. which revealed poor absorption and severe gastrointestinal effects [5]. The search for oral activity in platinum drugs thus became an area of chemical development distinct from other second and third-generation programmes. This led to the identification of a new class of platinum(IV) compounds with suitable properties for oral administration and high antitumour activity. The compound selected for clinical evaluation from this class, known as JM216, is shown in Fig. 3.

1.2
The Search for Oral Absorption

Early results compared the efficacy of cisplatin, carboplatin and tetraplatin (*trans*-d,l)-1,2diaminocyclohexane tetrachloroplatinum(IV) and selected second-generation analogues bearing a mixed amine (ammine/cyclohexylamine) carrier ligand when given by oral or intra-peritoneal routes to mice bearing the ADJ/PC6 plasmacytoma (Table 1). Results clearly demonstrated that, while the compounds are active when given orally, the doses required are significantly higher and that there is some reduction in the therapeutic index (the ratio of a toxic to a therapeutic dose).

Evidently, the low organic/aqueous partition coefficients for these compounds severely limit the level of absorption. The solubility in water and chloroform of several series of platinum complexes was reported by Tobe et al. in the 1970s [6]. Data for selected compounds are shown in Table 2. As the amine chain length increases there is a significant decrease in aqueous solubility but little change in chloroform solubility. Thus while the organic/aqueous distribution increases, favouring oral uptake, the severe reduction in aqueous solubility means that these compounds are very poorly absorbed. While it is not the case

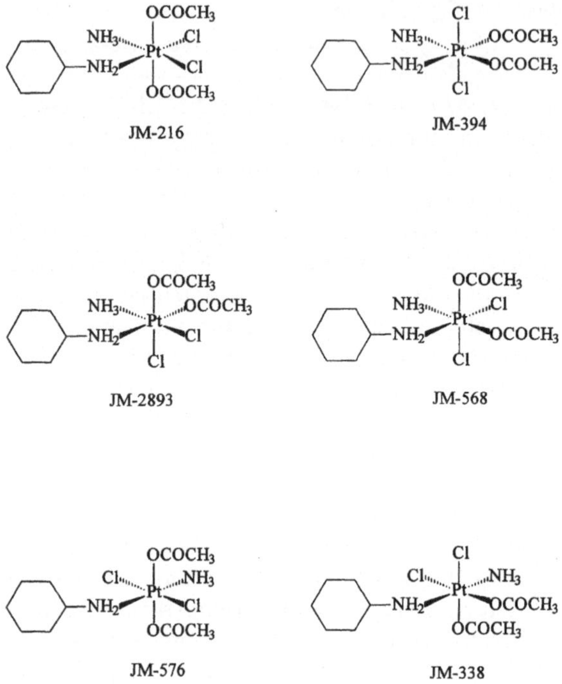

Fig. 3. Structures of JM216 and its isomers

Table 1. i.p. versus oral antitumour activity in mice bearing the ADJ/PC6 plasmacytoma of selected second-generation analogues

JM No.	i.p.		TI	p.o.		TI
	LD_{50} (mg/kg)	ED_{90} (mg/kg)		LD_{50} (mg/kg)	ED_{90} (mg/kg)	
JM118 (Pt II dichloro)	14	1	14	140	11	13
JM120 (Pt II malonate)	670	11.7	57	>1600	73	>22
JM149 (Pt IV dichloro dihydroxo)	17.4	0.4	44	118	18	7
JM132 (Pt IV tetrachloro)	29.5	0.8	37	120	22	5
Cisplatin	11.3	0.6	18.8	140	24	5.8
Carboplatin	180	14.5	12.4	235	99	2.4
Tetraplatin	22.5	0.8	28	480	45	10.6

General structure of Pt JM compounds

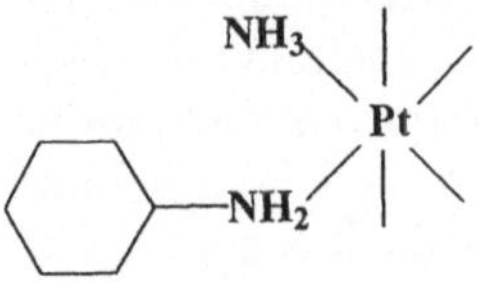

Table 2. Solubilities of platinum amine complexes

	Solubility		
	Aqueous (mM)	Chloroform (mM)	Ratio (aq/org)
cis-[PtCl$_2$(NH$_3$)$_2$]	8.9	<0.07	<0.008
cis-[PtCl$_2$(i-C$_3$H$_7$NH$_2$)$_2$]	0.22	0.00380	0.0038
cis-[PtCl$_2$(c-C$_4$H$_7$NH$_2$)$_2$]	0.021	0.0105	0.015
cis-[PtCl$_2$(n-C$_4$H$_9$NH$_2$)$_2$]	0.051	0.087	1.7
cis-[PtCl$_2$(c-C$_6$H$_{11}$NH$_2$)$_2$]	0.0041	0.0176	4.3
cis-[PtCl$_4$(i-C$_3$H$_7$NH$_2$)$_2$]	0.015	0.015	1.0
cis-[PtCl$_4$(c-C$_4$H$_7$NH$_2$)$_2$]	0.0019	0.0012	0.64
cis-[PtCl$_4$(n-C$_4$H$_9$NH$_2$)$_2$]	0.66	0.43	0.66
cis-[PtCl$_4$(c-C$_6$H$_{11}$NH$_2$)$_2$]	0.039	0.26	6.8

for all .compounds, comparison of the solubility data for platinum(II) complexes cis-[PtCl$_2$A$_2$] (A=amine) and platinum(IV) complexes cis-[PtCl$_4$A$_2$] indicates that a significant improvement in solubility can be achieved in both aqueous and organic media for the platinum(IV) series.

Since the amine ligands are an essential feature of the activity of platinum drugs, modification of these groups to promote oral absorption is likely to affect the activity of the complexes. Modification of the axial ligands of platinum(IV) complexes offers the prospect of achieving good absorption without strongly influencing activity, since it is believed that platinum(IV) complexes are reduced to platinum(II) compounds (involving the loss of the axial ligands) prior to reaction with DNA [7]. However, platinum(IV) compounds, as d$_6$ third row transition-metal complexes, are inert to substitution under most conditions and thus the synthesis of a suitable range of complexes presents some difficulties. These are overcome by exploiting the inertness of the metal-ligand bonds. Platinum(IV) dihydroxo complexes are well known and one such compound CHIP (iproplatin) was clinically tested as a potential second-generation drug [8]. The oxygen atom of the hydroxo ligand retains strong nucleophilic character after bonding to platinum, allowing it to participate in reactions with suitable electrophiles while the inertness of the bond to platinum ensures that the complex remains intact. This reaction has allowed the preparation of a class of complexes containing axial carboxylate ligands and greatly increased the scope for synthesising complexes with a wide range of partition coefficients and solubilities.

2
Preparation and Evaluation of Dicarboxylate Complexes

2.1
Synthesis of Platinum(IV) Dicarboxylate Complexes

Reaction of the platinum(IV) dihydroxo complexes with organic acids was generally unsuccessful in obtaining high yields of the desired product. However, reasonable yields (>50% isolated yield) were obtained for formate complexes. In other cases, conversion of the acid to an active anhydride (e.g. using isobutylchloroformate) or ester (e.g. using N-hydroxysuccinimide) generally allowed formation of the desired complex.

A reaction with much greater applicability was that of the platinum complex with an acid anhydride as electrophile. Stirring of the dihydroxo complex in the anhydride as solvent (or using an inert solvent such as hexane for solid anhydrides) at ambient temperature for several hours achieved essentially 100% conversion to the dicarboxylate complex and this reaction was used to obtain several series of products containing simple alkyl and aryl carboxylate ligands [9]. The systematic variation of these ligands could therefore be used to determine the optimum properties for selection of a candidate compound for evaluation as an orally administered drug.

Acid chlorides can be used as an alternative reagent to the anhydride although a suitable base such as triethylamine or pyridine must be added to remove the HCl which is formed as a byproduct of the reaction (USP 5072011, [10]).

Carboxylation can also be achieved using a variety of other electrophiles such as pyrocarbonates and isocyanates yielding carbonate and carbamate complexes, respectively [9].

As noted by Tobe [6], mixed amine complexes offer the prospect of improved solubility due to reduced lattice energies in the solid state when compared with the bis-amine series. While he discounted mixed amine complexes due to the difficulties of their preparation, improved methods of synthesis for K[Pt-Cl(NH$_3$)] made exploitation of these compounds possible. A wide range of mixed amine platinum(IV) carboxylate complexes have now been synthesised and evaluated.

The activity of diaminocyclohexane complexes of platinum has also attracted many groups of workers to synthesise compounds of this type. The preparation of platinum(IV) acetate and trifluoroacetate complexes from a range of platinum(II) oxalate and malonate complexes was reported by Khokhar and colleagues [11]. Derivatives of oxaliplatin [Pt(oxalato)(1,2-diaminocyclohexane)] containing carboxylate ligands ranging from acetate to octanoate were obtained by Kidani et al. [12]. In addition to preparing the bis-carboxylate complexes they also explored reactions to obtain the mono-carboxylate compounds. These were obtained either by reacting the bis-carboxylate with the appropriate acid or reacting the platinum(IV) *trans*-dichloro complex with silver carboxylate. In both cases chromatographic purification of the product was necessary [13].

2.2
Preclinical Studies of Dicarboxylates Leading to the Selection of JM216

Following the preparation of a large series of ammine/amine Pt(IV) dicarboxy-lates, a pivotal feature of the preclinical lead optimization process was a comparison of antitumour activity in mice bearing the ADJ/PC6 subcutaneous murine plasmacytoma by the oral versus intra-peritoneal routes of administration. The ADJ/PC6 tumour model has been widely used in platinum drug development programmes, including in the discovery of carboplatin [14], and is believed to predict well for activity in man. Confirmatory antitumour efficacy studies were performed in immune-suppressed mice bearing one of a selected group of subcutaneous human ovarian carcinoma xenografts [15]. These models exhibited a generally good correlation between responsiveness to platinum drugs and corresponding patient response data and in vitro cell line drug sensitivity [15,16].

It was soon established that, in contrast to the results shown in Table 1 for other ammine/amine Pt(II) and Pt(IV) complexes, the ammine/amine Pt(IV) dicarboxylate series generally exhibited no loss of antitumour activity (ED_{90}) by the oral route and substantially lower toxicities resulting in significant improvements in therapeutic indices (Table 3 [17,18]). Notably, some compounds such as JM274 and JM244 exhibited therapeutic indices by the oral route in this model of around 300. Ten dicarboxylates (JM251, JM216, JM225, JM269, JM221, JM223, JM274, JM244, JM291 and JM256; see Table 3 for structures) were then also evaluated by oral administration against a panel of five human ovarian carcinoma xenografts [19]. All compounds induced substantial antitumour growth delays (60 days or more) against the cisplatin-sensitive PXN/100 model while only JM244, JM216 and JM221 induced growth delays of 10 days or more in the cisplatin refractory SKOV-3 xenograft. Studies with three xenografts of intermediate sensitivity to cisplatin (HX/110, OVCAR-3, PXN/109 T/C) showed that JM216, JM225, JM269 and JM244 were the most active while JM274 and JM291 (the oxalate) were least active.

In parallel, a further level of selection involved a determination of oral absorption properties in rodents and emetogenic properties in the ferret. At this point, both JM244 and JM221 were shown to be highly emetogenic in the ferret compared to compounds in the acetate series (e.g. JM269) [18]. Comparative scores (duration of emesis in hours × mean number of episodes following a single oral dose) were 39 for JM244, 85 for JM221, 36 for cisplatin (i.v. administration) but only 4 for JM269, 9 for JM216 and 5 for carboplatin (i.v.). Hence, the acetato series was shown to be substantially less emetogenic than cisplatin and comparable to that observed for carboplatin. Extensive head-to-head studies of oral JM216, JM225 and JM269 versus i.v. cisplatin and carboplatin against four human ovarian carcinoma xenografts were then performed [19,20]. The acetato series produced broadly comparable tumour growth delays to that observed for i.v. cisplatin and carboplatin (Fig. 4). Further antitumour studies were conducted with four compounds (JM216, JM225, JM251 and JM269) using the M5076 murine sarcoma and A2780 human ovarian xenograft [21]. Oral antitumour ac-

Table 3. i.p. versus oral antitumour activity in mice bearing the ADJ/PC6 plasmacytoma of selected Pt(IV) dicarboxylates

JM No.	i.p.		TI	p.o.		TI
	LD_{50} (mg/kg)	ED_{90} (mg/kg)		LD_{50} (mg/kg)	ED_{90} (mg/kg)	
JM251; $R=cC_6H_{11}, R_1=H$	60	1.34	44.8	335	12	27.9
JM216; $R=cC_6H_{11}, R_1=CH_3$	30	5.7	5.3	330	5.8	56.9
JM225; $R=cC_5H_9, R_1=CH_3$	42	0.9	44.6	280	4.8	58.3
JM269; $R=cC_7H_{13}, R_1=CH_3$	71	11.7	6.1	>1600	10.4	>153.8
JM213; $R=nC_3H_7, R_1=C_3H_7$	17.5	5.4	3.2	140	5.2	26.9
JM221; $R=cC_6H_{11}, R_1=C_3H_7$	15.5	2.5	6.2	280	5.2	53.9
JM223; $R=iC_4H_9, R_1=C_3H_7$	17.5	3	5.9	240	3.1	77.4
JM274; $R=cC_6H_{11}, R_1=C_4H_9$	14	5	2.8	1120	3.6	311
JM244; $R=nC_3H_7, R_1=C_6H_5$	17.5	10.5	1.67	670	2.4	279
JM256; $R=cC_6H_{11}, R_1=NHC_2H_5$	41	2.9	14.1	237	3.9	60.8
JM291; $R=cC_6H_{11}, R_1=CH_3$	42	1.45	28.9	550	21.5	25.6

General structures

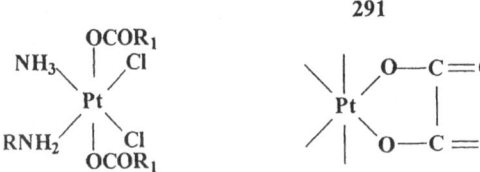

291

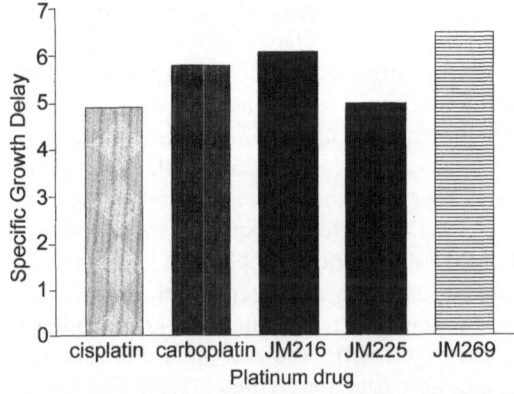

Fig. 4. In vivo antitumour activity against CH1 human ovarian xenograft

tivity was also observed in these two models, JM216 being the most active of the series against the M5076 sarcoma.

From the above-described antitumour and emesis studies, JM216 (Fig. 3) was selected on the basis that it possessed both good oral antitumour activity against a variety of tumour models, a low emesis score in the ferret and favourable physicochemical properties.

3
Chemistry of JM216

JM216 possesses a solubility in water of around 0.3 mg/ml, in saline 0.4 mg/ml and in 1-octanol of 0.7 mg/ml. The octanol/water partition coefficient was determined to be 0.1.

3.1
Crystal Structure

A crystal structure determination for JM216 revealed the expected octahedral coordination around platinum [22]. The N1-Pt-N2 angle (94.3°) is significantly greater than 90° with corresponding reductions in the N-Pt-Cl angles. As is common for platinum complexes containing nitrogen and oxygen donor ligands, there are a number of hydrogen bonds formed within the crystal. Intramolecular interactions occur between both nitrogen atoms and the carbonyl oxygen atoms. The orientation of the cyclohexylamine ligand towards one or other of the acetate ligands may serve as some protection from attack and produce some differentiation in the reactivity of these two groups.

3.2
Isomers

To confirm the specificity of the synthesis and the analytical procedures for the single isomer JM216, all other isomers of the complex $[PtCl_2(OAc)_2(NH_3)(c-C_6H_{11}NH_2)]$ were prepared [23]. While there are only *cis* and *trans* isomers for square planar platinum(II) complexes such as cisplatin, there are six isomers for these octahedral platinum compounds (see Fig. 3) with the two all-*cis* isomers also being optically active.

The all-*trans* dihydroxo-platinum(IV) complex may be prepared by oxidation of *trans*-$[PtCl_2(NH_3)(c-C_6H_{11}NH_2)]$ with aqueous hydrogen peroxide. On treatment with acetic anhydride in the dark, this complex undergoes a simple conversion to the corresponding all-*trans* bis-acetate. However, if the reaction is allowed to continue in light, then an isomerisation takes place to yield the isomer with *cis* acetate and *cis* chloride ligands, JM338. This is analogous to the isomerisation of *trans*-$[PtCl_2(OH)_2(NH_3)_2]$ studied by Sadler et al. [24]. Their studies suggest that the reaction may proceed via a reductive step involving loss of one of the hydroxo ligands. It seems likely that the mechanism for the acetato complexes is the same, with light-catalysed homolytic cleavage of one of the Pt-O or Pt-Cl bonds preceding the rearrangement.

Platinum(II) complexes containing *cis* acetate ligands are highly reactive with one acetate ligand being readily displaced [25]. Therefore such compounds are not good candidates for oxidation to obtain platinum(IV) complexes. However, it was discovered that non-aqueous oxidation of $[PtCl_2(NH_3)(c-C_6H_{11}NH_2)]$ could be used to obtain the desired compounds. The most efficient reagent was found to be the iodine(III) compound iodobenzene diacetate. Using *cis*-$[PtCl_2(NH_3)(c-C_6H_{11}NH_2)]$ as the starting material, while some intramolecular exchange reduced the yield of the reaction, the desired all-*cis* isomers JM568 and JM2893 (Fig. 3) could be isolated by chromatography. When starting with *trans*-$[PtCl_2(NH_3)(c-C_6H_{11}NH_2)]$ it was found that the reaction proceeded with rearrangement to give *cis* amine ligands producing JM394.

3.3
Stability

3.3.1
Acid and Alkali

Treatment of JM216 with acids, e.g. hydrochloric, results in protonation of the carboxylate groups and their subsequent substitution. However, the reaction is sufficiently slow (a half-life of several hours in 1 M HCl) to be of little consequence with regard to possible reactions in the stomach prior to absorption [13,17].

In contrast to this behaviour JM216 is unstable in alkaline media. While platinum(IV) complexes are usually relatively inert to substitution reactions, the amine complexes can undergo deprotonation equilibria which markedly alters their behaviour. This class of reactions has been most studied for cobalt(III) amine complexes and is generally believed to follow the mechanism described as D_{cb} (dissociative conjugate base mechanism). In the presence of strong base the amine ligands are deprotonated forming amido species. The *trans* effect of the amido ligand is much greater than that of the amine and so substitution of the ligand in the *trans* position becomes rapid. As JM216 contains two different amine ligands the differing equilibria for the ammonia and cyclohexylamine ligands result in differing substitution rates for the two chloride ligands. The cyclohexylamine protons are the more acidic which directs initial substitution *trans* to this ligand [23]. This reaction is of particular importance in the metabolism of JM216 (see Sect. 8).

3.3.2
Light

In common with many coordination complexes, JM216 is unstable to light. In solution, the degradation is complex involving both substitution and reduction.

Degradation in the solid state is very much slower, making a solid oral dosage form the most appropriate for complexes of this type.

3.3.3
Reduction

Since reduction of platinum(IV) to platinum(II) is required before reaction with DNA, it is important for JM216 to have a reduction rate which is an effective compromise between maintaining the platinum(IV) state for uptake and distribution, and reduction to platinum(II) sufficiently rapidly to achieve reaction with DNA rather than being excreted intact. A suitable model reaction for this reduction is that with ascorbate [17,26]. A mechanistic study of this reaction has recently been reported [27]. Studies by Kidani et al. [12] of oxaliplatin derivatives have shown that the reduction rate reduces as the carboxylate chain length increases. For JM216, a half-life for reduction of the order of 50 min in 5 mM ascorbate suggests an adequate lifetime for absorption as platinum(IV) species, but with the likelihood of reduction occurring in the body to form reactive metabolites (see Sect. 8).

4
In Vivo Antitumour Efficacy

In addition to the extensive oral antitumour studies described above in mice bearing the ADJ/PC6 plasmacytoma and human ovarian carcinoma xenografts, JM216 was also evaluated in vivo in two murine models of acquired cisplatin resistance and schedule-dependency effects were determined.

Although the demonstration of activity against acquired cisplatin-resistant tumour models was not a prerequisite for selection, JM216 did exhibit evidence of activity against the cisplatin-resistant variant of the ADJ/PC6 [20]. Following oral administration, a therapeutic index of 2.2 was achieved (ED_{90} of 180 mg/kg). JM216 was not particularly active against the murine L1210 ascitic leukaemia or its cisplatin-resistant subline. Maximum increases in life spans were 41% (i.p. against the parent tumour) and 21% (oral against the resistant tumour).

The schedule dependency of oral JM216 was studied using the ADJ/PC6 and the PXN/109 T/C ovarian carcinoma xenograft and compared single dose (every 21 days) versus once a day dosing for 5 consecutive days (every 21 or 28 days) versus once a day dosing indefinitely [28]. In contrast to cisplatin, daily × 5 administration of JM216 to mice bearing the ADJ/PC6 tumour improved the tolerance, the antitumour potency and the therapeutic index (i.e. TI of 56 single dose versus >423 daily × 5 JM216).

The schedule-dependency studies performed in the human ovarian carcinoma xenograft were especially important in helping to guide the clinical evaluation of the drug (see below). These studies clearly showed a gain (two- to threefold in terms of growth delay, $P<0.01$) in antitumour activity in the daily × 5 arm versus the single dose or chronic dosing schedules (Fig. 5). Moreover, greater to-

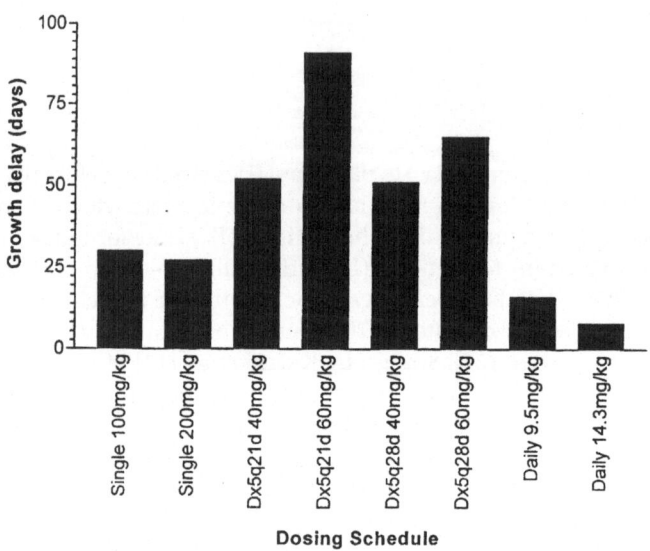

Fig. 5. JM216 oral schedule dependency

tal weekly doses were tolerable on the 5-d split dose schedule versus a single weekly dose (300 versus 200 mg/kg).

More recently, JM216 has been studied in vivo against the P388 leukaemia and M5076 sarcoma murine models in combination with orally administered etoposide [29]. When used in combination in non-tumour bearing animals, only about 25% of each drug's individual maximum tolerated dose could be safely used. Evidence of a therapeutic synergy was reported in the leukaemia model but not in the solid M5076 tumour. A phase I trial of this combination is ongoing in Canada.

5
In Vitro Antitumour and Cellular Properties

5.1
Cytotoxicity

Initially, JM216 was evaluated in vitro using the sulforhodamine B (SRB) assay against a panel of human ovarian carcinoma cell lines as part of an extensive series of platinum(II) and (IV) mixed amine complexes. The original panel was established to be representative of the range of clinical responsiveness to platinum-based chemotherapy observed in patients presenting with advanced ovarian cancer [30]. Greater than 100-fold range in intrinsic cellular sensitivity to cisplatin was observed. Many of the cell lines were also available as in vivo xenograft counterparts.

For the mixed amine platinum(IV) dicarboxylate series, a clear relationship existed between increasing the number of axial (R1, see Table 3 for generalised structure) ligand carbons and increasing cytotoxicity [31]. For variation of the amine ligand, changes in cytotoxicity were not as dramatic. Maximum cytotoxicity was observed with alicyclic substituents with increasing ring size from cyclobutane through to cycloheptane resulting in increasing cytotoxicity. Some of the longer chain dicarboxylates (i.e. butyrates and pentanoates of high lipophilicity) with an alicyclic R group were significantly more cytotoxic than cisplatin and probably are among the most cytotoxic platinum-containing molecules yet described. For example, JM274 (Table 3 for structure) was shown to be at least 100-fold more potent than cisplatin and moreover appeared to be selectively cytotoxic to the more intrinsically cisplatin-resistant cell lines such as HX/62 and SKOV-3 [31]. However, the predictability of these in vitro data into in vivo antitumour efficacy is known to be limited by rapid biotransformation as described for JM216 (see below).

JM216 (R=cyclohexane, acetato axial groups) showed a similar in vitro potency to that of cisplatin itself; mean IC_{50} values across the ovarian cell line panel of 1.7 μM versus 3.5 μM for cisplatin and 26.3 μM for carboplatin [20,31]. Spearman rank analysis of patterns of response across the panel was used to determine whether compounds were acting by similar (correlation coefficient approaching 1) or dissimilar means. Calculated correlation coefficients were: cisplatin/carboplatin 0.93 ($P<0.01$), cisplatin/JM216 0.86 ($P=0.01$), carboplatin/JM216 0.96 ($P<0.01$) and tetraplatin/JM216 0.46 ($P>0.05$). Thus JM216 in vitro appeared to behave similarly to cisplatin and carboplatin but not to the 1,2-diaminocyclohexane compound, tetraplatin. The reactivity of JM216 in comparison to that of cisplatin, carboplatin and tetraplatin was addressed in terms of the effect on cytotoxicity of differing drug exposure times [20]. While cytotoxicity was unchanged in moving from a 24- to 96-h exposure time for cisplatin and tetraplatin, for JM216, the IC_{50} was 1.8-fold lower than that obtained with a 24-h exposure [20].

JM216 was also shown to exert cytotoxic properties against panels of lung [32], murine L1210 [33] and cervix carcinoma [34] cell lines. As with the ovarian lines, Spearman rank analysis of the pattern of response in the small cell lung panel showed a high coefficient for the cisplatin/JM216 pair (0.82) [32]. Across a panel of five human cervix carcinoma cell lines, the potency of JM216 was similar to that of cisplatin with the exception of the HX/156 cell line which was 13-fold more sensitive to JM216 [34].

5.2
Circumvention of Acquired Cisplatin Resistance

Primarily as part of the evaluation process to identify a "third-generation" platinum complex possessing activity against cisplatin-resistant disease, models of acquired cisplatin resistance were established by in vitro exposure of the most sensitive ovarian cell lines (41M and CH1). Parallel mechanistic studies then determined the major underlying mechanisms of resistance in each of these pairs of lines. 41McisR was shown to be resistant primarily through reduced drug accumulation [35,36], whereas CH1cisR was shown to be resistant due to enhanced DNA repair/tolerance to platinum-DNA adducts [35]. In the original studies of dicarboxylates, the butyrates JM221 and JM244 (see Table 3 for structures) were shown to circumvent resistance in a cisplatin-resistant subline of the 41M ovarian carcinoma [31]. Perhaps, unsurprisingly, given the differences in lipophilicity between cisplatin and JM221/JM244, the dicarboxylates were shown to overcome acquired cisplatin resistance due to reduced drug uptake [35,36]. This "mechanism-related" circumvention of resistance was shown in the 41M/41McisR pairs of lines (resistance circumvention observed) versus the CH1/CH1cisR pair of lines (only partial circumvention of resistance observed).

The ability of JM216 to circumvent acquired cisplatin resistance in vitro has been addressed in a variety of pairs of lines (Fig. 6). In common with the longer chain dicarboxylates JM221 and JM244, JM216 also circumvented resistance in the transport-deficient 41McisR cell line [20,36]. A similar circumvention of acquired transport-mediated cisplatin resistance was shown for the HX/155 and HX/155cisR pair of cervix carcinoma cell lines [34]. Further transport studies using the 41M pair of lines revealed that the mechanism of JM216 transport across plasma membranes is through passive diffusion, predominantly as a result of its enhanced lipophilicity compared to cisplatin [37]. Circumvention of acquired cisplatin resistance by JM216 was also observed in the murine L1210 [33] and OVCAR-3 human ovarian carcinoma pairs of lines [20]. Partial circumvention of resistance in the CH1, A2780 ovarian, GCT27 testicular, and H69, MOR lung pairs of lines was also observed [20,32].

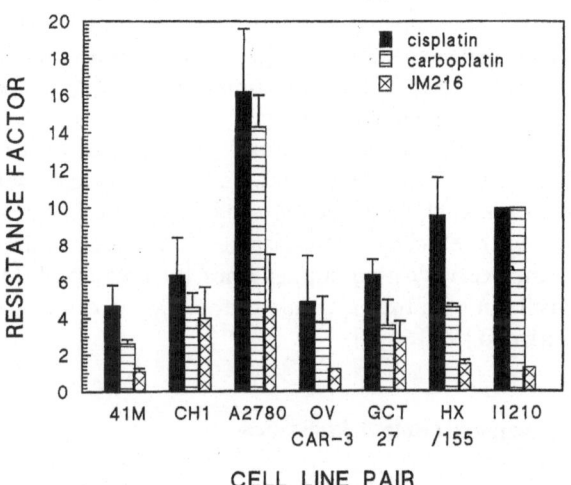

Fig. 6. In vitro cross-resistance profile

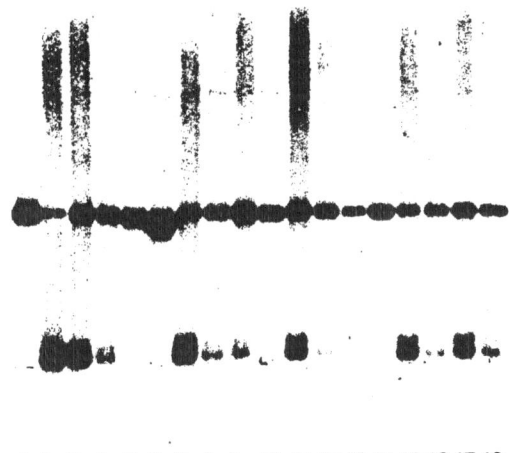

1 2 3 4 5 6 7 8 9 10 11 12 13 14 15 16 17 18

Lane 1: double strand control
Lane 2; single strand control
Lanes 3-6: 10 µM tetraplatin 1, 2, 4, 6 h exposure respectively
Lanes 7-10: 10 µM JM149 1, 2, 4, 6 h exposure respectively
Lanes 11-14: 10 µM JM216 1, 2, 4, 6 h exposure respectively
Lanes 15-18: 10 µM JM335 1, 2, 4, 6 h exposure respectively

Fig. 7. Interstrand crosslink formation by JM216 on naked DNA

5.3
DNA Binding Properties

The DNA binding properties of JM216 have been studied in terms of binding to naked DNA and within human ovarian carcinoma cell lines [38]. Exposure of plasmid DNA and analysis by agarose gel electrophoresis showed that, in common with cisplatin, JM216 (or metabolites thereof) was capable of forming platinum-DNA interstrand crosslinks (Fig. 7). This was confirmed within cells (CH1 and SKOV-3 ovarian carcinoma) using alkaline filter elution. The nature of platinum-DNA intrastrand crosslinks formed by JM216 was addressed using a competitive ELISA and the ICR-4 monoclonal antibody raised against cisplatinated DNA. DNA extracted from CH1 cells exposed to JM216 was recognised by ICR-4 but around twofold less effectively than adducts formed by cisplatin suggesting some differences in adduct recognition for the two drugs.

5.4
Acquired Resistance to JM216

Mechanisms of acquired resistance to JM216 were studied in two human ovarian carcinoma cell lines (41M and CH1) which had previously also been made resistant to cisplatin. Notably, in contrast to cisplatin (see above), 41MJM216R showed no deficiency in platinum transport [39]. Resistance to JM216 in the 41MJM216R cell line appeared to be due mainly to elevated glutathione (around 1.7-fold higher) reflected in a similar reduction in total platinum bound to DNA following JM216 exposure. These results suggest that, in contrast to cisplatin, acquired resistance to JM216 may be less likely to occur through reduced drug uptake although this has not yet been addressed in vivo. However, as shown in the JM216R cell lines, other resistance mechanisms common to cisplatin such as elevated glutathione (in 41MJM216R) and increased DNA repair (as observed in CH1JM216R) may also apply to JM216.

5.5
JM216 in Combination with Radiation

Recently, JM216 has also been studied in vitro in combination with ionising radiation in RIF1 mouse tumour cells [40]. While no radiosensitisation was observed with a 2-h drug exposure (irradiation occurring 15 min prior to the completion of exposure), significant radiosensitisation (1.5 enhancement ratio) was observed with 1- and 0.5-h exposures. These data suggest that JM216 could be used as a clinical alternative to cisplatin for combination with radiotherapy.

6
Preclinical Toxicology

6.1
Myelosuppression

JM216 showed limited toxicity. The dose-limiting toxicity in rodents was myelosuppression. In mice receiving a single dose of 200 mg/kg leucopaenia was the prominent effect with nadir at days 2 to 10 post-treatment and recovery was observed on day 14. Mild thrombocytopaenia and anaemia were also observed. When administered daily for 5 consecutive days at 55 mg/kg, thrombocytopaenia was the most significant effect with nadir reached by day 14 and recovery observed by day 30. Leucopaenia and anaemia were mild [41].

6.2
Nephrotoxicity

Comparative studies with cisplatin, carboplatin and JM216 at the maximum tolerated dose showed that JM216 is devoid of nephrotoxicity [42]. In mice at the maximum tolerated dose, cisplatin caused glycosurea, proteinurea and decreased the glomerular filtration rate after 4 days. These changes were not observed with carboplatin or JM216. In rats, at the maximum tolerated dose, while cisplatin caused a five-fold elevation of plasma creatinine and urea and decreased the creatinine clearance by ten-fold, neither JM216 nor carboplatin had any effect. There was also no sign of histological kidney toxicity after 4 days and no difference in kidney weight unlike the cisplatin-treated animals which showed renal tubular epithelial necrosis. The ^{14}C-inulin clearance examined 96 h postadministration of the maximum tolerated dose of carboplatin or JM216 was unaffected while a 45% reduction was observed after cisplatin (unpublished data).

6.3
Neurotoxicity

Neurotoxicity is a severe cumulative toxicity associated with cisplatin or tetraplatin treatment. In rats after 6 weeks of treatment with cisplatin twice weekly at 2 mg/kg, a significant decrease (17%) in sensory nerve conduction velocity was observed while the motor nerve conduction velocity was unaffected. Tetraplatin at 1 mg/kg twice weekly caused a 14% decrease in sensory nerve conduction velocity. Twice weekly treatment with JM216 at 25 mg/kg did not affect the sensory nerve conduction velocity [43].

6.4
Histopathology: Tests of Liver and Gut Functions

Histological abnormalities in mice at the maximum tolerated dose were confined to the intestinal tract. This was characterised by villus atrophy, crypt tip necrosis and reduced frequency of crypt mitosis similar in appearance and severity to that which was observed with cisplatin and carboplatin (unpublished data). However, there was no sign of mucosal damage as disaccharidase activity (sucrase, threhalase and maltase) was unaffected [41]. Both alkaline phosphatase and alanine aminotransferase were unaffected indicating no liver damage.

6.5
Emesis

As mentioned earlier, the emetogenic properties of JM216 were evaluated by Bristol Myers Squibb in the ferret. Ferrets of 1.5 kg in weight received an oral bolus of JM216 in suspension in arachis oil or water with 0.1% tween 80. For comparative evaluation, carboplatin and cisplatin were administered i.v. at equitoxic doses. The duration of the emetic response induced by JM216 was significantly shorter than that induced by cisplatin (Bristol Myers Squibb). In mice emesis is not observed but stomach bloating has been described as a good predictor of emesis. The stomach bloating observed following 200 mg/kg of JM216 was significantly lower than that observed following the equivalent dose of cisplatin and comparable to that observed after carboplatin (Fig. 8).

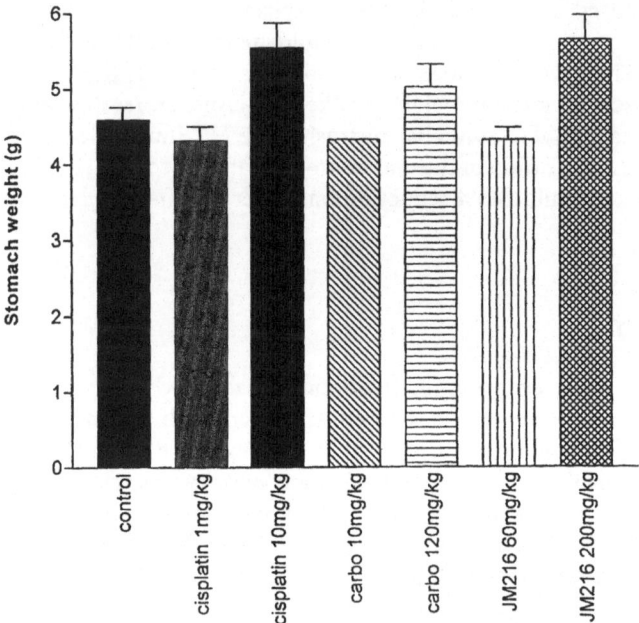

Fig. 8. Stomach bloating assay in mice

7
Preclinical Pharmacology

7.1
Plasma Pharmacokinetics in Balb C⁻ Mice

The pharmacokinetics of total and free platinum administered orally to Balb C⁻ mice at doses of 20, 50, 100 and 200 mg/kg is presented in Table 4.

Table 4. Summary of pharmacokinetic parameters in mice (values in italics are ultrafilterable platinum)

Dose (mg/kg)	20	50	100	200
C_{max} (µg Pt/ml)	1.3±0.38	4.94±1.2	4.58±0.9	10.4±0.21
	0.416	*0.76±0.12*	*1.97*	*1.4±0.21*
T_{max} (min)	60	30	60	120
	30	*30*	*30*	*30*
$t_{1/2\beta}$ (h)	29.5	29.7	31.4	31.6
$t_{1/2\beta}$ (min)	*135*	*87*	*108*	*126*
AUC_{0-72h} (µg Pt/ml h^{-1})	31.4	95	106	270
	1.07	*1.4*	*1.5*	*4.4*
$AUC_{0-\infty}$ (µg Pt/ml h^{-1})	44.2	113	172	338
	1.23	*1.43*	*3.05*	*4.99*

The maximum platinum levels in plasma were observed between 30 and 120 min and were delayed with increasing dose. Platinum unbound to proteins was detectable up to 7 h postdosing. Elimination of total platinum was biphasic with a terminal half-life of ~30 h. The half-life for free platinum ranged between 89–135 min which is considerably longer than the 10 min reported for cisplatin or 25 min reported for carboplatin [44]. The relationship between the different pharmacokinetic parameters and dose showed that C_{max} for free platinum does not vary linearly with dose (r=0.603) while the $AUC_{0-\infty}$ for both free and total platinum shows a linear increase with dose (r=0.996, 0.990) (Table 4, unpublished observations). However in another study where sampling was only performed up to 8 h at doses of 9.5, 40 and 200 mg/kg, it was found that AUC_{0-8h} and C_{max} for total and ultrafiltrable platinum increased in a non-linear fashion with increasing dose [41].

The pharmacokinetic parameters were unaffected by repeated administration [41].

7.2
Tissue Distribution

Forty-eight hours after 200 mg/kg oral administration of JM216 to Balb C^- mice platinum levels were the highest in the liver (6–19 µg Pt/g tissue) and kidney (2.8–12 µg Pt/g tissue). This is five times higher than that which has been reported after equivalent doses of cisplatin. All other tissues (spleen, heart, lung) had levels ≤3.1 µg Pt/g tissue. In the liver a time course of platinum levels showed that the C_{max} were reached by 2 h postadministration [42].

7.3
Excretion

Following administration of 200 mg/kg JM216 orally(in oil or in saline) 8% of platinum was eliminated in urine over 72 h and 66% was present in the faeces after 72 h (unpublished observations).

8
Metabolism

The metabolism of JM216 is complex, leading to at least six metabolites of different antitumour potency. Initially, metabolic studies were performed in fresh human plasma incubated with JM216 and ^{15}N-JM216. The incubated plasma was then chromatographed by HPLC followed by atomic absorption spectrophotometry or LC/MS to evaluate the platinum-containing fractions. This led to the identification of JM383, JM518, JM559 and JM118 (Fig. 9) [45]. Selected ion monitoring revealed that these species are also present in patient plasma ultrafiltrate at peak concentration following treatment with JM216 [46] and that JM118 is the main metabolite of JM216 [47]. No parent drug could be detected

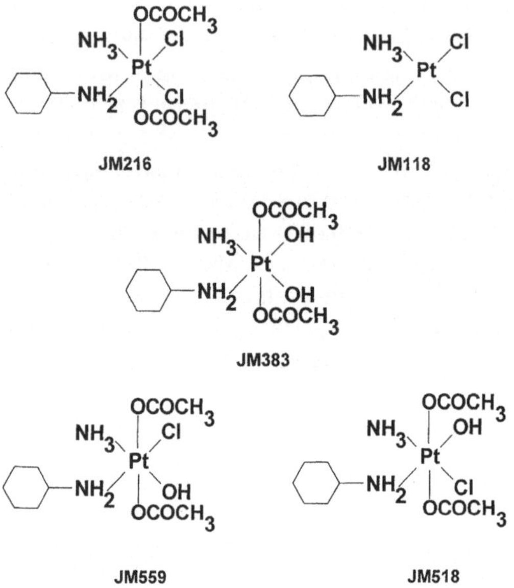

Fig. 9. Metabolites of JM216

in any of the patient samples examined even as early as 15 min postadministration. A further platinum-containing fraction could be detected in patients plasma ultrafiltrate but not in ultrafiltrate from animals treated with JM216, JM118, JM383 or JM518. [47,48]. This unidentified metabolite was not observed either in tumour cells treated with JM216. An early eluting platinum-containing fraction could be detected in all matrices evaluated (patient artificial plasma incubations, animal plasma, ovarian carcinoma cells). This peak was shown to be more important in cells with high glutathione levels and decreased if cells were treated with buthionine sulfoximine (which inhibits glutathione synthesis) suggesting that it contains a glutathione adduct [49]. The activity of JM118, JM518 and JM383 against our panel of ovarian carcinoma cell lines proved to be of the same order of magnitude as that of the parent compound while the glutathione adduct corresponds to a detoxification product [48]. In conclusion, as expected, the axial ligands were readily lost leading to the main metabolite JM118. However, surprisingly, ligand-exchange reactions with replacement of a Cl atom by a hydroxyl group was observed up to 4 h post-treatment in JM216 (see discussion, Sect. 3.3). Conjugation with glutathione once again proved to be the main detoxification pathway (see Fig. 10 for summary).

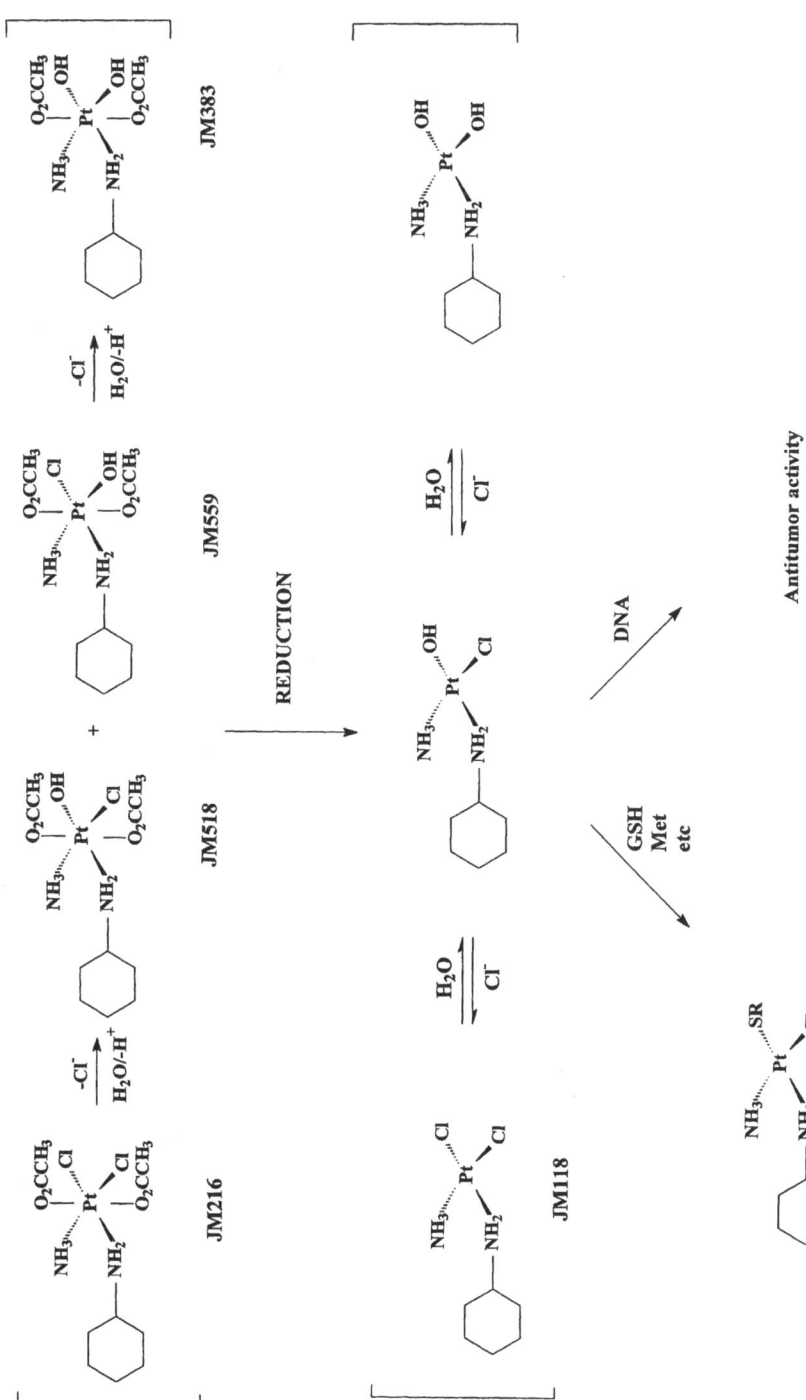

Fig. 10. JM216 Biotransformation pathways

9
Clinical Trials

9.1
Phase I Single Dose Study

JM216 entered phase I clinical trials in August 1992 as a single oral dose (60–700 mg/m^2) delivered every 21 days using a dry filled gelatin capsule. No hydration or diuresis was associated with the treatment. As predicted by the preclinical studies, there was no sign of neurotoxicity or nephrotoxicity. The emesis was easily controllable by prophylactic anti-emetics. However, due to limited absorption/dissolution of the drug, no dose-limiting toxicity was observed in this trial. One of the 37 patients showed a partial response in relapsed ovarian carcinoma following cisplatin treatment. Two other patients showed a significant decrease in tumour markers (CA125) [50].

9.2
Phase I Daily × 5 Study

Having failed to reach MTD with a single dose, the second phase I study used a daily×5 schedule in 32 patients at doses of 20, 30, 60, 100 and 140 mg/m^2/day [51]. Prophylactic anti-emetics were used routinely (ondansetron +dexamethasone). Under these conditions, JM216 was well tolerated with good control of emesis. The MTD was 140 mg/m^2 with 2 of 3 patients experiencing grade 4 thrombocytopaenia and grade 3 and 4 leucopaenia. There was no neurotoxicity, no ototoxicity and no renal toxicity observed. No objective sign of response was recorded in this trial. Although considerable interpatient variability was observed for a given dose, the pharmacokinetic parameters (AUC, C$_{max}$) in this study proved to increase linearly with dose for both total and ultrafiltrable Pt (e.g. r=0.78 for free AUC). The maximum concentrations for free platinum in the plasma were reached around 2 h and the half-life for free platinum varied from 4 to 14 h. There were no differences in the pharmacokinetic parameters observed on day 1 and day 5. A sigmoidal relationship was observed between the plasma ultrafiltrate AUC and the severity of thrombocytopaenia (r=0.83). A limited sampling strategy was developed and showed that sampling at 2, 4 and 6 h postadministration could predict the AUC for free platinum with a 93% predictability [52]. The doses recommended for phase II studies were 100 mg/m^2 in patients previously treated with platinum-based chemotherapy and 120 mg/m^2 otherwise.

9.3
Phase I Twice Daily Study

Nineteen patients received 150 to 350 mg/m^2 JM216 orally twice daily. Again, the considerable variability in the pharmacokinetic parameters (mainly absorption) in both plasma and plasma ultrafiltrate stopped the trial before MTD was reached. No response was recorded but two patients with mesothelioma had stable disease [53].

9.4
Phase II Studies

JM216 has been evaluated in ovarian cancer and prostate cancer. Although the results are presently unpublished, a phase II study is planned in hormone refractory prostate cancer against and combined with prednisolone. In non-small-cell lung cancer no objective response was recorded [54]. A study is planned to evaluate JM216 as a radiosensitiser.

In summary, the clinical properties of JM216 have been well predicted by the preclinical studies in rodents. The dose schedule dependency and the toxicology profile have been confirmed and the pharmacokinetic properties shown to be very similar to those observed in rodents.

10
Summary

JM216 was designed to improve upon the poor gastrointestinal absorption properties of cisplatin and carboplatin and thus allow oral platinum-based chemotherapy in patients. This was achieved through the synthesis of a new class of ammine/amine Pt(IV) complexes possessing lipophilic axial carboxylate ligands. A large number of this class of complex exhibited oral antitumour activity in the ADJ/PC6 plasmacytoma model. JM216 was selected because it possesses both good oral antitumour activity, including effectiveness against a panel of human ovarian carcinoma xenografts, favourable toxicological properties in rodents and relatively low emetogenic properties in the ferret model. JM216 entered phase I clinical trials at the Royal Marsden Hospital, London as the first orally available platinum drug. As predicted by the preclinical studies, myelosuppression (both thrombocytopenia and neutropenia) was dose-limiting. Saturable absorption leading to non-linear pharmacokinetics was observed in the single dose phase I resulting in daily schedules being introduced, each schedule repeated for 5 days. Phase II trials in a variety of tumour types are ongoing. Circumvention of acquired cisplatin resistance was observed in vitro especially in resistant lines possessing a deficiency in cisplatin transport; it is too early to judge whether this observation is of clinical relevance in patients with cisplatin refractory disease.

Acknowledgements. The drug development programme leading to the discovery of JM216 involved a collaboration between Johnson Matthey (Sonning, UK and West Chester, Pennsylvania), the Drug Development Section/CRC Centre for Cancer Therapeutics at the Institute of Cancer Research and Bristol Myers Squibb. Numerous colleagues/collaborators of ours have contributed including chemists at Johnson Matthey (Barry Murrer, Mike Abrams, Chris Giandomenico), biologists at ICR under the directorship of Ken Harrap (Prakash Mistry, Sarah Morgan, Mervyn Jones, Phyllis Goddard, George Abel, Frances Boxall, Swee Sharp, Ciaran O'Neill, Melanie Valenti, Grace Poon), clinicians at ICR/Royal Marsden Hospital (Mark McKeage, Ian Judson) and scientists at BMS (Anna Marie Casazza, Bill Rose, John Schurig, Al Crosswell). Thanks are also due to Swee Sharp for the preparation of some of the figures.

References

1. Harrap KR (1985) Cancer Treat Rev 12:21
2. Kelland LR (1993) Crit Rev Oncol/Hematol 15:191
3. Gore ME, Fryatt E, Wiltshaw E, Dawson T, Robinson BA, Calvert AH (1989) Br J Cancer 60:767
4. Cleare MJ, Hoeschele JD (1973) Bioinorg Chem 2:187
5. Van Hennik MB, van der Vijgh WJF, Klein I, Vermorken JB, Pinedo HM (1989) Cancer Chemother Pharmacol 23:126
6. Tobe ML, Khokhar AR (1977) J Clin Haematol Oncol 7:114
7. Lempers ELM, Reedijk J (1991) Adv Inorg Chem 37:175
8. Barnard CFJ, Cleare MJ, Hydes PC (1986) Chem Br 22:1001
9. Giandomenico CM, Abrams MJ, Murrer BA, Vollano JF, Rheinheimer MI, Wyer SB, Bossard GE, Higgins JD (1995) Inorg Chem 34:1015
10. Galanski M, Keppler BK (1996) Inorg Chem 35:1709
11. Al-Baker S, Siddik ZH, Khokhar AR (1994) J Coord Chem 31:109
12. Kizu R, Nakanishi T, Miyazaki M, Tashiro T, Noji M, Matsuzawa A, Eriguchi M, Takeda Y, Akiyama N, Kidani Y (1996) Anti-cancer Drugs 7:248
13. Kidani Y, Kizu R, Miyazaki M, Noji M, Matsuzawa A, Takeda Y, Akiyama N, Eriguchi M (1996) In: Pinedo HM, Schornagel JH (eds) Platinum and other metal coordination complexes in cancer chemotherapy 2. Plenum Press, New York, p 43
14. Goddard PM, Valenti MR, Harrap KR (1991) Annals Oncol 2:535
15. Harrap KR, Jones M, Siracky J, Pollard L, Kelland LR (1990) Ann Oncol 1:65
16. Kelland LR, Jones M, Abel G, Harrap KR (1992) Cancer Chemother Pharmacol 30:43
17. Giandomenico CM, Abrams MJ, Murrer BA, Vollano JF, Barnard CFJ, Harrap KR, Goddard PM, Kelland LR, Morgan SE (1991) Synthesis and reactions of a new class of orally active Pt(IV) antitumour complexes. In: Howell SB (ed) Platinum and other metal coordination complexes in cancer chemotherapy. Plenum Press, New York, p 93
18. Harrap KR, Murrer BA, Giandomenico C, Morgan SE, Kelland LR, Jones M, Goddard PM, Schurig J (1991) Ammine/amine platinum IV dicarboxylates: a novel class of complexes which circumvent intrinsic cisplatin resistance. In: Howell SB (ed) Platinum and other metal coordination complexes in cancer chemotherapy. Plenum Press, New York, p 391
19. Kelland LR, Jones M, Gwynne JJ, Valenti M, Murrer BA, Barnard CFJ, Vollano JF, Giandomenico CM, Abrams MJ, Harrap KR (1993) Int J Oncol 2:1043
20. Kelland LR, Abel G, McKeage MJ, Jones M, Goddard PM, Valenti M, Murrer BA, Harrap KR (1993) Cancer Res 53:2581
21. Rose WC, Crosswell AR, Schurig JE, Casazza AM (1993) Cancer Chemother Pharmacol 32:197
22. Neidle S, Snook CF, Murrer BA, Barnard CFJ (1995) Acta Cryst C51:822
23. Barnard CFJ, Vollano JF, Chaloner PA, Dewa SZ (1996) Inorg Chem 35:3280
24. Kuroda R, Neidle S, Ismail IM, Sadler PJ (1983) Inorg Chem 22:3680
25. Canovese L, Tobe ML, Annibale G, Cattalini L (1986) J Chem Soc Dalton Trans 1107

26. Blatter EE, Vollano JF, Krishna BS, Dabrowiak JC (1984) Biochemistry 23:4817
27. Bose RN, Weaver EL (1997) J Chem Soc Dalton Trans 1797
28. McKeage MJ, Kelland LR, Boxall FE, Valenti MR, Jones M, Goddard PM, Gwynne J, Harrap KR (1994) Cancer Res 54:4118
29. Rose WC (1997) Cancer Chemother Pharmacol 40:51
30. Hills CA, Kelland LR, Abel G, Siracky J, Wilson AP, Harrap KR (1989) Br J Cancer 59:527
31. Kelland LR, Murrer BA, Abel G, Giandomenico CM, Mistry P, Harrap KR (1992) Cancer Res 52:822
32. Twentyman PR, Wright KA, Mistry P, Kelland LR, Murrer BA (1992) Cancer Res 52:5674
33. Orr RM, O'Neill CF, Nicolson MC, Barnard CFJ, Murrer BA, Giandomenico CM, Vollano JF, Harrap KR (1994) Br J Cancer 70:415
34. Mellish KJ, Kelland LR, Harrap KR (1993) Br J Cancer 68:240
35. Kelland LR, Mistry P, Abel G, Loh SY, O'Neill CF, Murrer BA, Harrap KR (1992) Cancer Res 52:3857
36. Loh SY, Mistry P, Kelland LR, Abel G, Harrap KR (1992) Br J Cancer 66:1109
37. Sharp SY, Rogers P, Kelland LR (1995) Clin Cancer Res
38. Mellish KJ, Barnard CFJ, Murrer BA, Kelland LR (1995) Int J Cancer 62:717
39. Mellish KJ, Kelland LR (1994) Cancer Res 54:6194
40. Van de Vaart PJM, Klaren HM, Hofland I, Begg AC (1997) Int J Radiat Biol Phys 72:675
41. McKeage MJ, Morgan SE, Boxall FE, Murrer BA, Hard GC, Harrap KR (1994) Cancer Chemother Pharmacol 33:497
42. McKeage MJ, Morgan SE, Boxall FE, Murrer BA, Hard GC, Harrap KR (1993) Br J Cancer 67:996
43. McKeage MJ, Boxall FE, Jones M, Harrap KR (1994) Cancer Res 54:629
44. Siddik ZH, Jones M, Boxall FE, Harrap KR (1988) Cancer Chemother Pharmacol 21:19
45. Poon GK, Mistry P, Raynaud FI, Harrap KR, Murrer BA, Barnard CFJ J Pharmaceut Biomed Analysis 13:1493
46. Poon GK, Raynaud FI, Mistry P, Odell DE, Kelland LR, Harrap KR, Barnard CFJ, Murrer BA (1995) J Chromatog 712:61
47. Raynaud FI, Mistry P, Donaghue A, Poon GK, Kelland LR, Barnard CFJ, Murrer BA, Harrap KR (1996) Cancer Chemother Pharmacol 38:155
48. Raynaud, FI, Boxall FE, Goddard P, Barnard CF, Murrer BA, Kelland LR (1996) Anticancer Res 16:1857
49. Raynaud FI, Odell DE, Kelland LR (1996) Br J Cancer 74:380
50. McKeage MJ, Mistry P, Ward J, Boxall FE, Loh S, O'Neill C, Ellis P, Kelland LR, Morgan SE, Murrer B, Santabarbara P, Harrap KR, Judson IR (1995) Cancer Chemother Pharmacol 36:451
51. McKeage MJ, Raynaud F, Ward J, Berry C, Odell D, Kelland LR, Murrer B, Santabarbara P, Harrap KR, Judson IR (1997) J Clin Oncol 15:269
52. Raynaud FI, Heybrook W, Judson IR (1997) Br J Cancer, 75:22
53. Beale P, Raynaud F, Hanwell J, Berry C, Moore S, Odell D, Judson I (1998) Cancer Chemother Pharmacol (in press)
54. Judson IR, Cerny T, Epelbaum R, Dunlop D, Smyth J, Schaefer B, Roelvink M, Kaplan S, Hanauske A (1997) Ann Oncol 8:604

trans-Platinum Compounds in Cancer Therapy: A Largely Unexplored Strategy for Identifying Novel Antitumor Platinum Drugs

Giovanni Natile[1], Mauro Coluccia[2]

[1] Dipartimento Farmaco-Chimico, Università di Bari, via E. Orabona 4, 70125 Bari, Italy
[2] Dipartimento di Scienze Biomediche e Oncologia Umana, Università di Bari, piazza G. Cesare, 70126 Bari, Italy
E-mail: [1] *natile@farmchim.uniba.it*, [2] *mauro.coluccia@dimo.uniba.it*

In recent years, several exceptions to the rule that the presence of two good leaving groups in *cis* positions of platinum complexes is a necessary condition for their antitumor activity have been reported. These "exceptions", which frequently show activity against *cis*-DDP-resistant cell lines, fall into three classes: 1) platinum(IV) complexes with the general formula *trans*-[PtCl$_2$X$_2$(L)(L')] (X=hydroxide or carboxylate; L, L'=ammine or amine), 2) *trans*-[PtCl$_2$(L)(L')] with L and/or L'=pyridine-like ligands, and 3) *trans*-[PtCl$_2$L$_2$] with L=iminoether. Greater inertness in biological media appears to be a common feature of these compounds. Both inter-strand cross-links and single-strand breaks have been proposed as cytotoxic lesions for platinum(IV) species, which might require reduction to platinum(II) prior to their interaction with DNA. Increased cell-membrane permeability, binding affinity for alternating puzine-pyrimidine sites, and enhanced inter-strand cross-linking ability were found for the second class of compounds. Finally, stable monofunctional adducts with duplex DNA, capable of inhibiting DNA and RNA synthesis, appeared to be the cytotoxic lesion for platinum-iminoether complexes. Taken together, these results show that *trans*-geometry does not impose steric constraints that inhibit antitumor activity and indicate the possibility of identifying novel antitumor platinum drugs in a largely unexplored area.

Keywords. Platinum antitumor drugs, *trans*-Configuration, Pyridine-like ligands, Iminoether ligands, Mixed ammine/amine ligands

1
Introduction

cis-[PtCl$_2$(NH$_3$)$_2$] (*cis*-DDP or cisplatin) is unusual among modern pharmaceuticals in that it is based on a metallic element, whereas most drugs are purely organic. The serendipitous discovery resulted from an experiment designed by a biophysicist, Barnett Rosenberg, to examine the effect of an electric field (produced via platinum electrodes in an ammonium chloride-containing nutrient solution) on the growth of bacteria [1]. A platinum electrode electrolysis product, *cis*-[PtCl$_4$(NH$_3$)$_2$], induced filamentous growth of the bacteria [2, 3]. This

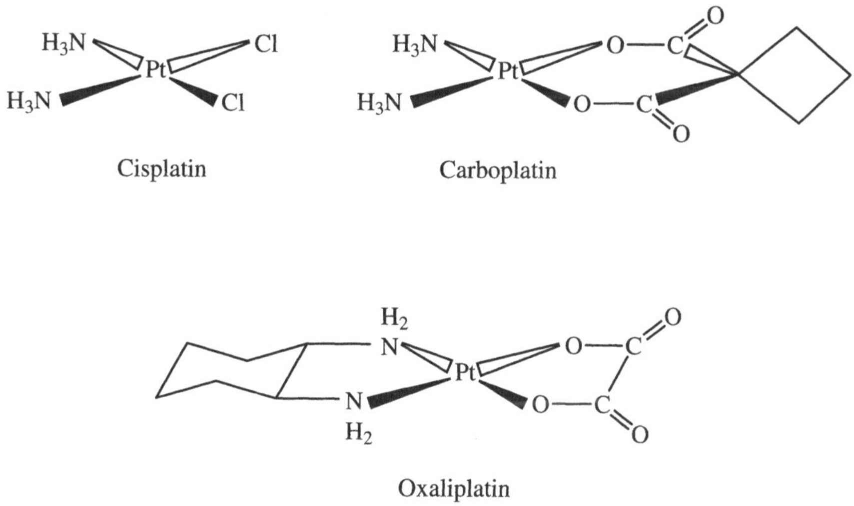

Cisplatin   Carboplatin

Oxaliplatin

Scheme 1

molecule was synthesized by standard methods and proved to be responsible for inhibition of bacterial cell division. *Cis*-DDP (Scheme 1) was found among the related species that were examined for activity [3].

The greatest impact of *cis*-DDP has been on the treatment of testicular [4, 5] and ovarian cancers [6], where its inclusion – in combination chemotherapy regimens – frequently produces cures in the former case and substantially improves survival in the latter. Additionally, the drug has proved to be beneficial in the treatment of head and neck, lung, and bladder cancers.

The nephrotoxicity and neurotoxicity associated with *cis*-DDP treatment have given impetus to the development of platinum analogs.

Substitution of the more stable cyclobutanedicarboxylate for the two chlorides led to the new drug carboplatin [7]. This drug diminishes renal effects [8], produces substantially less nausea, vomiting, and neurotoxicity, and finds dose-limiting toxicity in myelosuppression [9].

Manipulation of the structure of the leaving group appears to influence tissue and intra-cellular distribution of the platinum coordination complexes, but it is unlikely to prevent cross-resistance. It was hypothesized that modification of the carrier ligands, so that the analog produces a different spectrum of DNA lesions, might circumvent this problem.

One of the most successful *cis*-DDP analogs containing chelate amine carrier ligands is oxaliplatin [Pt(DACH)(oxalate)] (DACH=1,2-diaminocyclohexane), which was successfully developed in France [10–12]. The DNA adducts of oxaliplatin are predominantly intra-strand cross-links, as with *cis*-DDP and carboplatin [13, 14]. Evidence for non-cross-resistance has been obtained in vivo [11], and its potential role in the initial treatment of colorectal cancer is being investigated [15–17].

1.1
Significance of the *cis* Configuration for Anticancer Activity

Much of the current understanding of the mechanism of action of platinum drugs comes from studies with *cis*-DDP. The initial study by Rosenberg and colleagues [1] demonstrated that *cis*-DDP induced filamentous growth in bacteria while inhibiting cell division. This suggests that *cis*-DDP interferes mainly with DNA replication rather than with RNA and protein synthesis [18]. Interaction with DNA is believed to occur after *cis*-DDP loses its chloride ligands through aquation to yield a reactive electrophile [19]. This situation occurs in biological systems when chloride concentration is low, as in the intra-cellular medium. Thus the term of *leaving group* has been applied to the chloride ligands or – in the case of analogs – to the moieties that replace chloride ligands and are displaced under physiological conditions. In contrast, the amino groups, as well as similar substituents that are not susceptible to displacement in physiological environments, are termed *carrier ligands*.

From this early beginning, the *cis* configuration was identified as potentially critical for activity because the *trans* isomer, *trans*-[PtCl$_2$(NH$_3$)$_2$] (*trans*-DDP or transplatin), was a far less potent inhibitor of bacterial cell division. It has been speculated that the 1,2-intra-strand adducts [d(GpG)Pt and d(ApG)Pt], which produce a local kinking and unwinding of duplex DNA, are responsible for the cytotoxic effect of *cis*-DDP, accounting for 65 and 20% respectively of the total bifunctional adducts) (Scheme 2) [20, 21]. The *trans* isomer, which does not form 1,2-intra-strand lesions, exhibits comparatively little antitumor activity

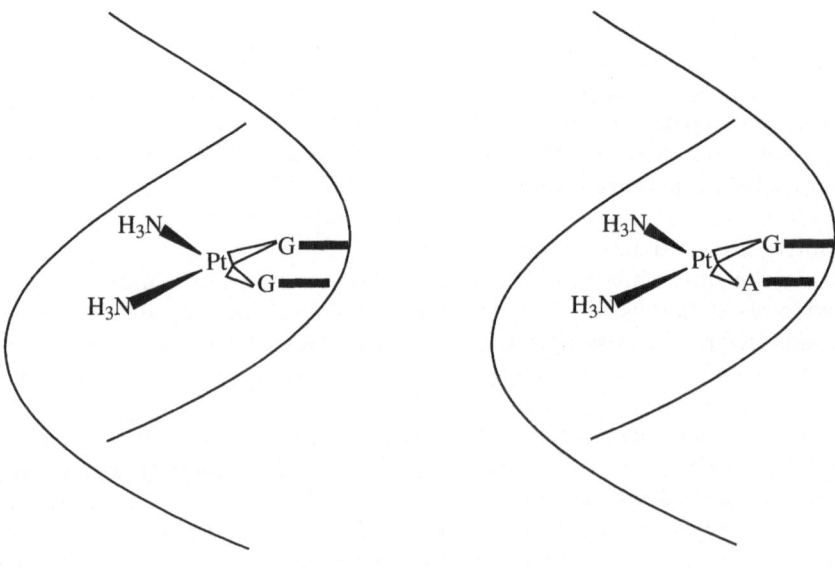

Scheme 2

[22]. Therefore the significance of the *cis* configuration to anticancer activity has nearly achieved the status of dogma, and has guided most of the subsequent work on the development of platinum analogs, modified either in the leaving groups or in the carrier ligands.

In recent years, several exceptions to the rule that the presence of two good leaving groups in the *cis* positions of platinum complexes is a necessary condition for antitumor activity have been reported. These "exceptions", which frequently show activity against *cis*-DDP-resistant cell lines, fall into three classes: 1) platinum(IV) complexes with the general formula *trans*-$[PtCl_2X_2(L)(L')]$ (X= hydroxylide or carboxylate; L, L'=ammine, amine) [23, 24], 2) *trans*-$[PtCl_2(L)(L')]$ with L and/or L'=pyridine-like ligands [25–29], and 3) *trans*-$[PtCl_2L_2]$ with L=iminoether [30–32]. Bis-platinum(II) complexes with bridging diamine ligands of composition *trans*-$[\{PtCl(NH_3)_2\}_2(\mu\text{-}NH_2(CH_2)_xNH_2)]Cl_2$ [33, 34] have not been included in this chapter, because they have leaving groups on two different platinum centers (linked together by a bridging ligand); these cannot be regarded as *trans* in the classical sense: two groups bound to the same metal center on opposite sides.

2
Activation of the *trans* Geometry by Platinum(IV) Species

2.1
Exploration of the Platinum(IV) Oxidation State for *cis*-Oriented Complexes

The platinum(IV) oxidation state has already been explored for *cis*-oriented complexes (Scheme 3). In particular, after promising initial results, the complexes *cis,trans,cis*-$[PtCl_2(OH)_2(iPrNH_2)_2]$ (iproplatin) and $[PtCl_4(DACH)]$ (DACH=1,2-diaminocyclohexane, tetraplatin or ormaplatin) proved not to be useful. In the case of the iproplatin, testing in phase II trials failed to reveal activity. In the case of tetraplatin, the platinum(IV) conversion to a platinum(II) metabolite occurs within minutes [35]. However *cis,trans,cis*-$[PtCl_2(CH_3COO)_2(NH_3)(cC_6H_{11}NH_2)]$ (JM216), the lead compound in a series of mixed ammine/amine dicarboxylates, was the first orally active platinum compound to be discovered [7]. It is well adsorbed after oral administration and is taken up by cells resistant to *cis*-DDP on the basis of transport [36, 37]. Myelosuppression represents the dose-limiting toxicity [38].

There is compelling evidence to suggest that these platinum(IV) complexes (all possessing *cis*-amine carrier ligands) exert their antitumor effect through binding to DNA, to produce a spectrum of monofunctional and bifunctional adducts similar to that of their platinum(II) counterpart (both intra- and interstrand) [39, 40].

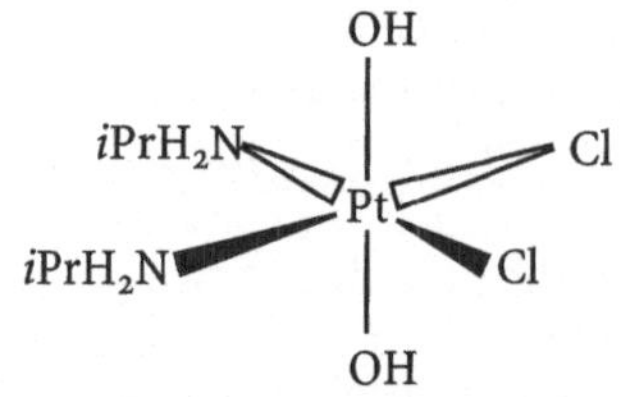

iproplatin 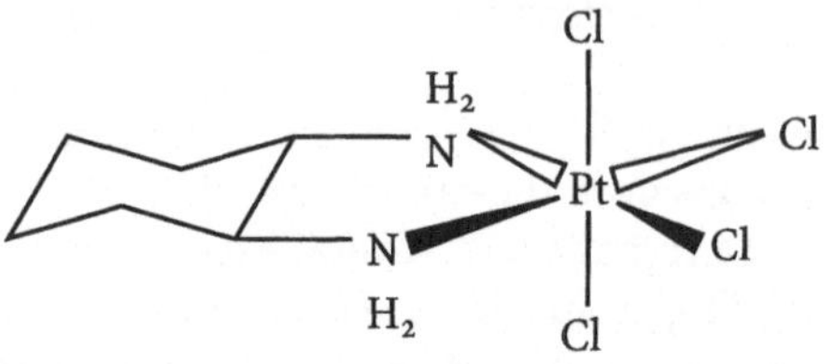   tetraplatin or ormaplatin

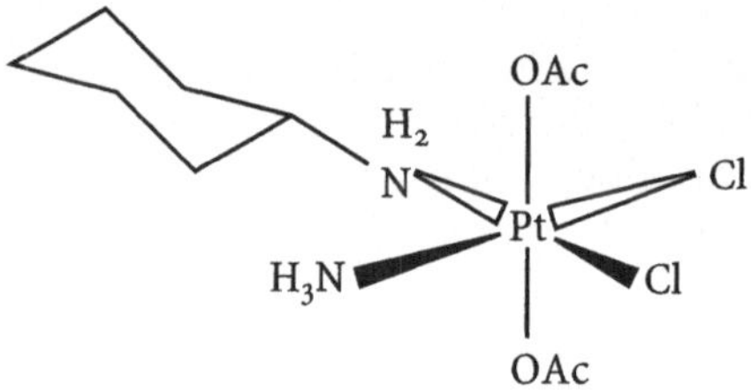

JM216

Scheme 3

2.2
Preparation of *trans*-Oriented Platinum(II) and Platinum(IV) Species

The platinum(IV) oxidation state proved to be very useful for conferring antitumor activity on platinum complexes in which the leaving ligands have *trans* geometry [23]. Within a platinum drug-discovery program performed over many years to identify more effective platinum-based anticancer drugs, with the particular aim of circumventing resistance to *cis*-DDP, a series of over forty compounds with *trans* geometry of the carrier ligands were investigated, half of them platinum(II) and half their platinum(IV) counterparts [41].

The synthesis of the *trans* isomers of platinum(II) is usually accomplished by the method originally proposed by Peyrone [42] and later modified by Kauffman and Cowan [43], i.e. by action of concentrated HCl on the aqueous solution of [Pt(amine)$_4$]Cl$_2$ species. The method utilizes the difference in the *trans* effect of halide and amine ligands in platinum(II) complexes to achieve selective substitution and thus to control the stereochemistry (Scheme 4).

The oxidation of platinum(II) complexes with hydrogen peroxide yields platinum(IV) complexes in which the stereochemistry of the platinum(II) complex is retained and *trans* hydroxo ligands are added [44–46]; chlorine reacts in an analogous manner. Dicarboxylate and dicarbamate complexes can be obtained from the dihydroxo species by reaction with carboxylic anhydride and alkylisocyanate, respectively [47].

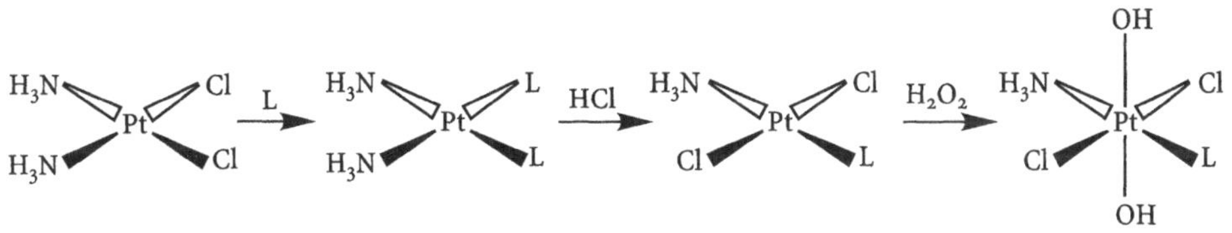

Scheme 4

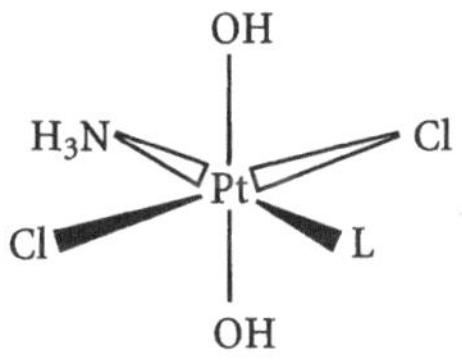

L =

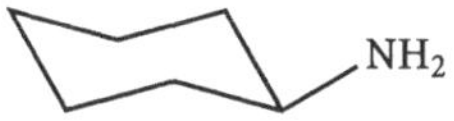

18

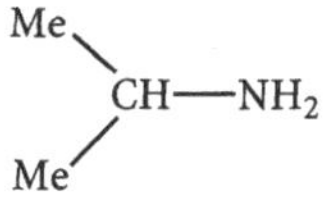

21

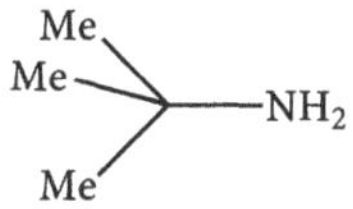

22

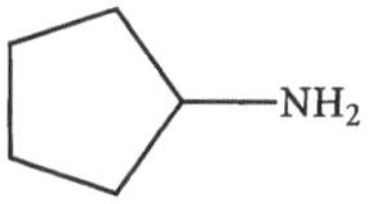

25

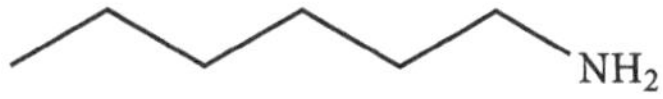

27

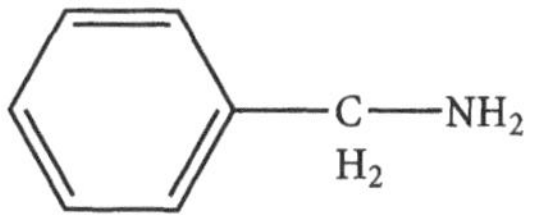

28

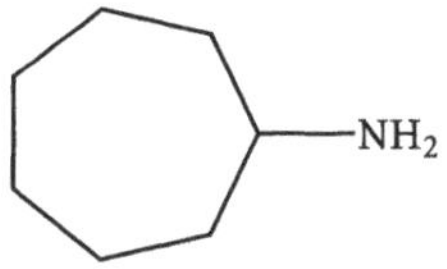

29

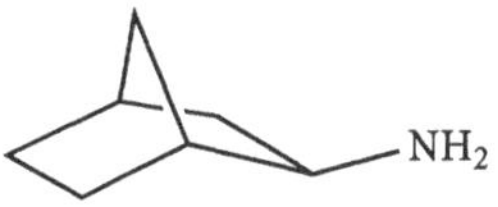

30

32

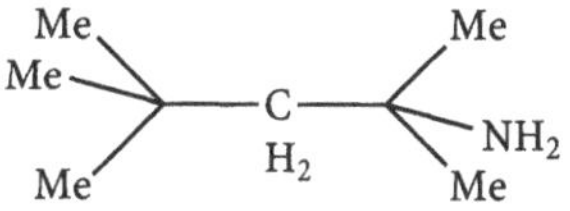

33

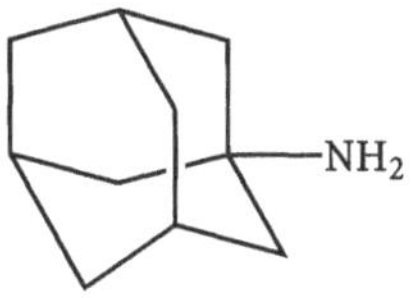

34

The same as **18** but with a second cyclohexylamine in place of NH$_3$ 35

The same as **18** but with Br in place of Cl   36

The same as **18** but with ethylcarbamate in place of OH   41

Scheme 5

2.3
In Vitro and in Vivo Testing

Many of the *trans* complexes, studied in vitro against a panel of human cell lines, exhibited comparable potency to *cis*-DDP and also overcame acquired resistance, which was mainly due to either reduced uptake or to enhanced removal of the platinum-DNA adduct [41]. Fourteen *trans* complexes showed significant *in vivo* antitumor activity against a subcutaneous murine ADJ/PC6 plasmacytoma model. All of them were platinum(IV) complexes with axial hydroxide ligands; only one possessed axial ethylcarbamate ligands (Scheme 5). When tested, all of their respective dichloroplatinum(II) or tetrachloroplatinum(IV) counterparts were inactive. Three out of the fourteen complexes (**18, 29,** and **34**) retained some efficacy against a *cis*-DDP resistant variant of the ADJ/PC6 plasmacytoma and five (**18, 22, 27, 28,** and **36**) exhibited antitumor activity against subcutaneously grown advanced-stage human ovarian carcinoma xenografts.

2.4
Stability in Tissue-Culture Media

Preliminary data indicated that the marked differences in in vivo antitumor effects may relate to their respective stability in tissue-culture growth media at 37 °C. For instance half times (h) for the disappearance of parent complexes were:
- *cis*-DDP-like species ca. 4.5,
- *trans*-DDP-like species ca. 1–2,
- *cis,trans,cis*-[PtCl$_2$(CH$_3$COO)$_2$(NH$_3$)(cC$_6$H$_{11}$NH$_2$)] (JM149) >96, and
- *trans,trans,trans*-[PtCl$_2$(OH)$_2$(NH$_3$)(cC$_6$H$_{11}$NH$_2$)] (JM335) ca. 24 (Scheme 6) [41].

Thus the inactivity of *trans* complexes of platinum(II) may be related to their chemical instability, which precludes the delivery of the active species to the tumor site. The additional stability of the dihydroxylate-platinum(IV) counterpart (such as JM335) appears to be sufficient to confer in vivo antitumor effects against a variety of tumor models. Therefore increased inertness of the platinum(IV) species could be the key factor for antitumor activity of *trans*-platinum species.

2.5
Cytotoxicity Related to Cross-Links and Single-Strand Breaks

In a comparative study of the binding properties of five *cis*-oriented compounds (*cis*-DDP, tetraplatin, JM118, JM216, and JM149) and two *trans*-oriented compounds (*trans*-DDP and JM335) (Schemes 3 and 6) in two human ovarian carcinoma cells (SKOV-3 and CH1), no correlation was found between levels of total platinum bound to DNA and cytotoxicity [24]. For the *cis*-oriented compounds,

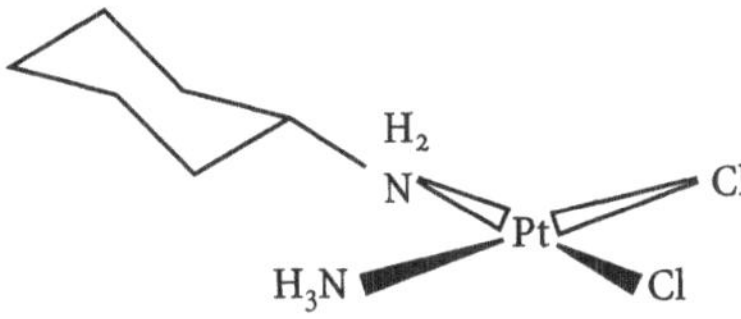

Scheme 6

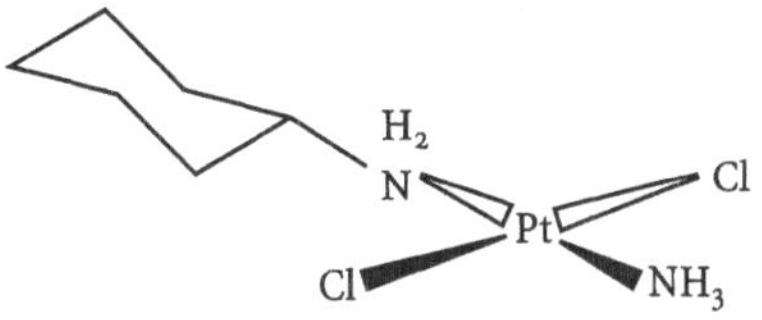

a strong positive correlation was found between cytotoxicity and recognition by ICR4 (monoclonal antibody raised against DNA platinated by *cis*-DDP). JM335 was efficient in forming inter-strand cross-links in SKOV-3 but, notably, none was observed in CH1 cells. In CH1 cells – in contrast – single-strand breaks were observed with JM335. It also appears that these platinum(IV) complexes require reduction prior to DNA binding [48], however direct interaction between Pt^{IV} and DNA cannot be excluded [49].

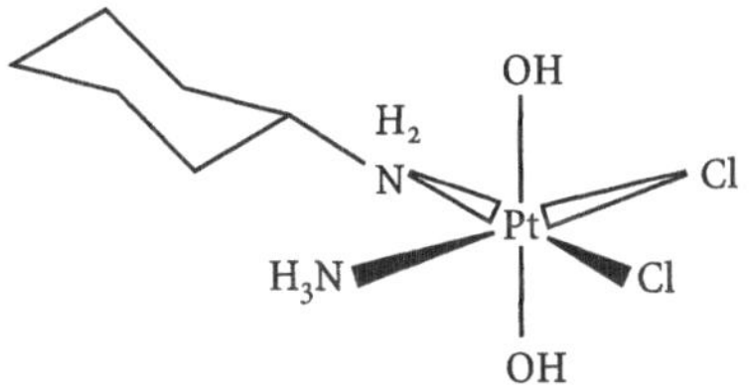

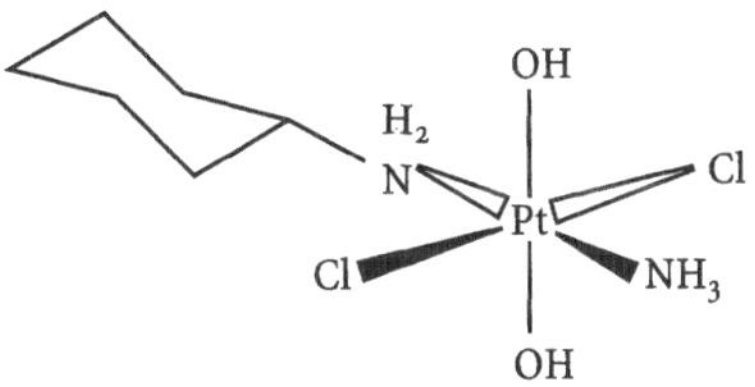

JM335, the most widely studied *trans*-oriented platinum(IV) species, has cytotoxicity comparable to that of *cis*-DDP itself and over fifty-fold greater than *trans*-DDP. Thus, IC$_{50}$ (the drug concentration which inhibits 50% cell growth) for JM335 is 3.1 µM, as compared to 4.1 µM for *cis*-DDP and 162 µM for *trans*-DDP. Moreover, the fact that JM335 shows no cross-resistance, whether resistance is mediated mainly by reduced platinum accumulation or by either enhanced removal of Pt-DNA adducts or increased tolerance to them[23], represents a structural lead to the development of platinum complexes exhibiting non-cross-resistance to *cis*-DDP [50].

3
Activation of the *trans* Geometry by Pyridine-Like Carrier Ligands

3.1
Preparation of *trans*-Complexes with Pyridine-Like Ligands

Activation of the *trans*-platinum geometry using sterically hindered ligands was first reported by Farrell [25]. Cytotoxicity of the *trans* structure equivalent to that of the analogous *cis*-isomer and of *cis*-DDP itself appeared to be a general feature of complexes with planar ligands of formula *trans*-[PtCl$_2$(L)(L')] [26]. Three distinct series of complexes were examined: (a) L=L'=pyridine, N-methylimidazole, or thiazole, (b) L=quinoline and L'=RR'SO (R= Me, R'=Me, Bz, or Ph), and (c) L=quinoline and L'=NH$_3$ (Scheme 7).

The method of Kaufman [37], exemplified in Scheme 4, was also used for the preparation of *trans*-[PtCl$_2$(L)(L')] complexes (L, L'=pyridine or N-methylimidazole [51]). In the case of thiazole the conversion of [Pt(thiazole)$_4$]Cl$_2$ to *trans*-[PtCl$_2$(thiazole)$_2$] was accomplished by heating the starting platinum-tetrathiazole dichloride salt at 100 °C instead of treatment with HCl [26]. The *trans*-[PtCl$_2$(RR'SO)(quinoline)] complexes were prepared by direct reaction of a sus-

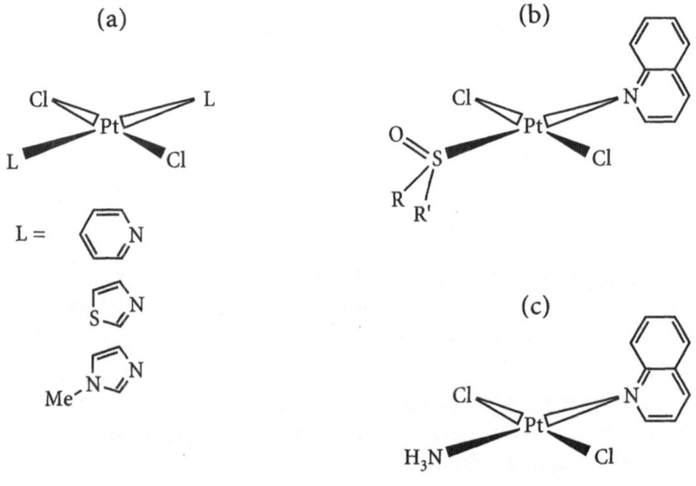

(a)   (b)

(c)

Scheme 7

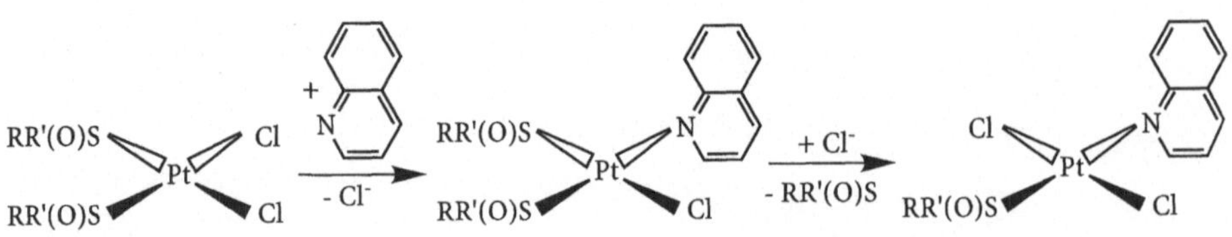

Scheme 8

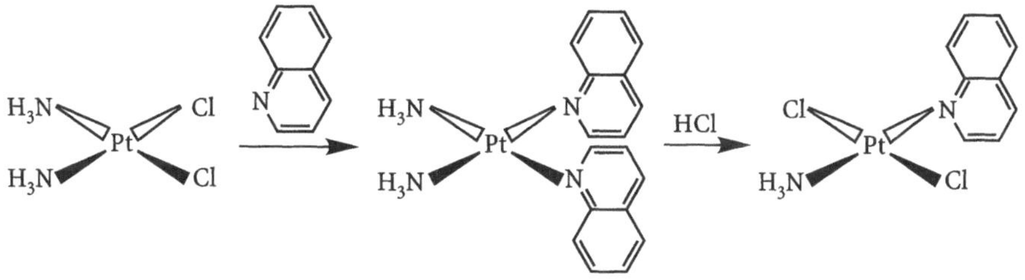

Scheme 9

pension of *cis*-[PtCl$_2$(RR'SO)$_2$] complexes with quinoline [26]. The reaction course is presumably that depicted in Scheme 8 [52].

The *trans*-[PtCl$_2$(NH$_3$)(quinoline)] complex was prepared according to the method of Kauffman and Cowan [43] summarized in Scheme 9.

3.2
Cytotoxicity of the *trans* Structure Compared to That of the *cis* Isomer and of *cis*-DDP

In general, the cytotoxicity of all *trans* complexes was approximately one-order of magnitude greater than that of *trans*-DDP; furthermore, the *trans* complexes were at least as cytotoxic as their direct *cis*-analogs [27]. Despite the apparent structural dissimilarity between the three series of complexes, a similar cytotoxicity profile was found using the COMPARE program [53]. This was different from that of *cis*-DDP but similar to that of the antibiotic rifamycin, that acts in the cell through inhibition of DNA-dependent RNA polymerases [54]. Finally the range of cytotoxicity of all *trans*-platinum complexes, defined as IC$_{50}$ (least sensitive)/IC$_{50}$ (most sensitive), was found to be significantly smaller for all *trans*-platinum complexes than for *cis*-DDP [27].

As a general feature, *trans* complexes with bulky planar ligands were not cross-resistant to *cis*-DDP (resistance factor=1) in both murine and human tumor cell lines resistant to *cis*-DDP.

Although high antitumor activity in vivo was not achieved with these complexes [26], the observed enhanced cytotoxicity of these *trans*-platinum complexes is nevertheless important mechanistically, because it demonstrates that there is no restriction, per se, to the development of active antitumor *trans*-platinum species.

3.3
Cellular Uptake and Chemical Reactivity

Uptake appeared to be greater for the pyridine- as compared to the ammine complexes (L1210 cells) and was greater for the *trans*- over the *cis*-isomer by a factor of ca. 4. Binding of the pyridine complexes to calf thymus DNA was almost equivalent for the *cis*- and *trans*-isomers but significantly less than for the analogous ammine complexes; the order of binding affinity was *trans*-DDP>>*cis*-DDP>>*cis*-[PtCl$_2$(py)$_2$]>*trans*-[PtCl$_2$(py)$_2$]) [27].

The planar ligands appear to hinder approach of incoming nucleophiles to the axial positions of the platinum center sterically. This steric feature may have some consequences with respect to reactivity, not only with DNA but also with glutathione (γ-L-glutamyl-L-cysteinylglycine, GSH) and other sulfur containing biomolecules. *trans*-DDP is more reactive than *cis*-DDP with GSH, but the presence of pyridine ligands slows down the platinum-GSH reaction.

3.4
DNA-Binding and Unwinding

The sequence specificity of *trans*-[PtCl$_2$(py)$_2$] includes alternating purine-pyrimidine sequences [28], in contrast the major stop sites for the *cis*-pyridine complex correspond to sites for *cis*-DDP [55]. The alternating purine-pyrimidine sequences would be expected to be a good source of inter-strand cross-links; in fact, *trans*-[PtCl$_2$(py)$_2$] showed a much greater frequency of inter-strand cross-links than *cis*-[PtCl$_2$(py)$_2$] (14–23% cross-linked material as compared to <5% [28]) at the same value of r_b (drug to nucleotide ratio), although it was much less reactive in binding to DNA [56]. Inter-strand cross-links could persist even in *cis*-DDP resistant cells, due to a higher proportion of inter-strand cross-links or to the formation of cross-links that are structurally different from those of *cis*-DDP.

Pyridine substitution for ammine also results in inversion of the *cis/trans* structure-activity relationship that had been previously observed for DNA-unwinding ability [57]. Unwinding angle (Φ, deg) is in the order *trans*-[PtCl$_2$(py)$_2$] (17°) >*cis*-DDP (13°) >*trans*-DDP (9°) >*cis*-[PtCl$_2$(py)$_2$] (4°) [28].

4
Activation of the *trans* Geometry by Iminoether Carrier Ligands

4.1
Iminoether Ligands: Preparation and Isomerism

A third class of antitumor-active *trans*-platinum complexes, reported by the group of Coluccia and Natile, involves iminoethers as carrier ligands; they have general formula *trans*-[PtCl$_2$(iminoether)$_2$] [iminoether=HN=C(OR')R] [30–32]. The X-ray structure of one of these complexes is shown in Fig. 1 [58].

Iminoethers are N-donor ligands that share several features with both aliphatic and aromatic amines. Like aromatic amines, iminoether ligands are planar, the hybridization of the nitrogen atom being sp^2 rather than sp^3 in aliphatic amines. However, like aliphatic amines, they have at least one hydrogen atom linked to nitrogen that is suitable for hydrogen-bond formation. Iminoethers are also asymmetric ligands, which are sterically hindered on one side

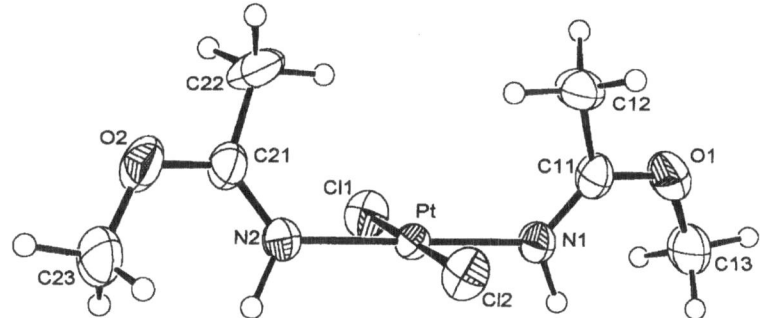

Fig. 1. Stereo view of *trans*-[PtCl$_2${*E*-HN=C(OMe)Me}$_2$] (*trans-EE*): drawn according to the atomic coordinates reported in Ref. 58

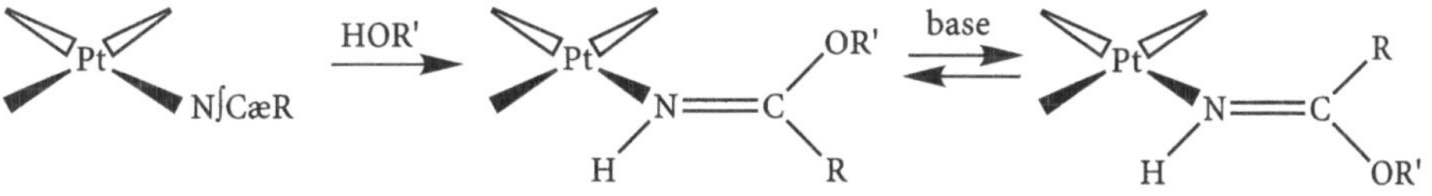

Scheme 10

of the nitrogen atom: that opposite the proton. Another peculiarity of imi-noethers is the possibility of isomerism within the ligand moiety. Depending upon the relative position of the substituents with respect to the C=N double bond, *E* and *Z* configurations can be obtained, corresponding to having the platinum and the alkoxy group *trans* and *cis* with respect to the C=N double bond, respectively [58].

As in the case of *cis*- and *trans*-DDP, the N-donor ligands of iminoether complexes are non-leaving groups in water at physiological pH, and are retained in adduct formation with biological substrates. Hydrolysis involves the chloride ligands exclusively, leading initially to the formation of a monoaqua species. Further transformations then lead to a mixture of several species, in which, however, the geometry of the complex (*cis* or *trans*) and the configuration of the iminoether ligands (*E* or *Z*) are always retained.

Iminoether ligands are generated in situ by addition of an alcohol to a nitrile that is already coordinated to platinum. The addition reaction takes place readily in the presence of base and leads to the formation of the *Z*-isomer, i.e. the alkoxide and the hydrogen add to the triple bond from opposite sides. Subsequently, *Z*→*E* isomerization can take place under basic conditions. The relative thermodynamic stability of the *Z* and *E* isomers depends upon the size of the alkoxide (OR') and the R group of the nitrile. The *Z* isomer is thermodynamically favored for small OR' and large R groups – vice versa for the *E* isomer (Scheme 10).

4.2
Cytotoxicity in Human Tumor Cell Lines

Inhibitory activity to in vitro cell growth was evaluated in comparison to that of
cis-DDP in a panel of human tumor cell lines containing examples of ovary
(A2780), colon (LoVo), lung (Calu-3), and breast (MDA, SKBR3) cancers, as well
as in a subline of ovarian cancer cells with acquired resistance to cis-DDP
(A2780/cp8) [59]. The trans-iminoether compounds showed a growth inhibitory potency similar to that of cis-DDP (mean IC_{50}=8 µM), and were able to circumvent cis-DDP resistance of A2780/cp8 cells, showing a resistance factor [IC_{50}
(acquired-resistance cells)/IC_{50} (sensitive cells)] of ca. 1.

Significantly, excision-deficient cells were four times as sensitive to trans-iminoether complexes as normal cells and, similarly, the drug was much more toxic
towards a strain of Drosophila defective in excision repair systems, thus impli-
cating cellular DNA as the cytotoxic target.

4.3
Antitumor Activity in Mice

The antitumor activity (P388 leukemia-bearing mice) of the cis-iminoether
complexes did not depend upon the treatment schedule (% T/C=145, where
T/C=life span of treated animals over that of control animals), and a cross-resistance of the same compound with cis-DDP was observed on the P388/DDP subline. In contrast, the trans-iminoether isomers were more active when administered daily for 7 consecutive days (% T/C=196) and also had a significant effect
on the cis-DDP-resistant subline (% T/C=133). These results indicate that substitution of the iminoether for ammines brings about only slight changes in the
activity of the cis complexes, whereas it has a dramatic effect on the behavior of
the trans species [30]. The same trans-iminoether complex showed an activity
comparable to that of cis-DDP in reducing the primary tumor mass and lung
metastases in mice bearing Lewis lung carcinoma; it thus represents a trans-plat-
inum complex active on both limphoproliferative and solid metastasizing
murine tumors [31]. This was the first unambiguous evidence for antitumor activity of a trans-platinum(II) complex.

4.4
Cellular Uptake and Chemical Reactivity

Like trans-[$PtCl_2(py)_2$] [27], trans-iminoether complexes exhibit greater cell
permeability and slower reactivity towards DNA (in cell-free media) than trans-
DDP. However, in contrast to the case of trans-[$PtCl_2(py)_2$], in which DNA reaches saturation at a much lower level of platination, the level of binding reached in
the case of trans-iminoether complexes after 14 h reaction time at 37 °C is comparable to that of ammine species [31]. Therefore, it appears that the iminoether
ligands, like pyridine, slow down the reaction with a sterically demanding nu-

cleophile, such as a nucleobase inserted in a DNA duplex, probably as a result of steric hindrance. However, unlike pyridine, iminoether ligands do not prevent reaching of the same level of platination as ammine compounds. This difference could result from the iminoether ligand having the steric bulk only on one side of the nitrogen donor.

4.5
Binding to DNA

4.5.1
Monofunctional Versus Bifunctional Adduct Formation

trans-Iminoether complexes interact preferentially with guanine residues, but, unlike *cis*-DDP – which blocks DNA synthesis by interacting at multiple guanine sites – the iminoether complexes also show affinity for isolated guanines in py-G-py sequences [59, 30].

In double helical (calf thymus) DNA, monofunctional adducts are preferentially formed at guanine residues even when DNA is incubated with the platinum complex for a relatively long time (48 h at 37 °C in 10 mM $NaClO_4$). In the case of *trans*-DDP under similar conditions, formation of bifunctional DNA adducts is predominant [60].

The ability to form inter-strand cross-links in calf thymus DNA was also investigated, since antitumor activity stemming from this type of lesion cannot be excluded [61, 62]. Under strictly analogous experimental conditions, *trans*-iminoether complexes showed greatly reduced DNA inter-strand cross-linking ability (heat denaturation/renaturation assay) compared to either *trans*-DDP, or to *cis*-DDP and *cis*-iminoether complexes. This appears to be another striking difference between *trans*-iminoether complexes and *trans*-$[PtCl_2(py)_2]$, the latter species manifesting an unusually high ability to form inter-strand cross-links: up to 14–23% of total platinum bound.

The above observations were also confirmed on plasmid DNA (pGEM-4Z [59], pBR322 [30], and pSP73 [63]). In agreement with the observation that *trans*-iminoether complexes form kinetically stable monofunctional adducts in double-helical DNA [59, 60], the measured unwinding angle was only 6°, a value that is identical to unwinding angles induced by monofunctional dienPt (dien= diethylenetriamine) and $Pt(NH_3)_3$. Greater unwinding angles are induced by bifunctional *cis*-DDP (13°) [64, 65] and *trans*-DDP (9°) [57] and also by *trans*-$[PtCl_2(py)_2]$ (17°) [28]. The work on plasmid DNA confirmed the low tendency of platinum-iminoether complexes to form inter-strand cross-links (<3%, compared to ca. 6 and 12%, respectively, for *cis*- and *trans*-DDP). It also confirmed the low rate of formation of such cross-links ($t_{1/2}$ >18 h, compared to ca. 4 and 11 h for *cis*- and *trans*-DDP, respectively) [66].

4.5.2
Non-Denaturational Conformational Distortion

Double-strand DNA modified by *trans*-[PtCl$_2${*E*-HN=C(OMe)Me}$_2$] (*trans-EE*) did not inhibit the binding of antibodies which either specifically recognize two neighboring purines of the same strand of DNA *cis*-coordinated to the platinum atom of a Pt(amine)$_2^{2+}$ moiety (Ab$_{cis}$) [67, 68]), or those that specifically recognize short single-strand segments in a double-strand DNA containing a *trans*-Pt(ammine)$_2^{2+}$ moiety (Ab$_{trans}$). In contrast double-strand DNA modified by *trans-EE* and subsequently thermally denatured, or DNA which was thermally denatured before modification with *trans-EE*, inhibited the binding of Ab$_{trans}$ to their immunogens with an efficiency similar to that of DNA modified by *trans*-DDP. Thus, *trans*-iminoether complexes induces conformational changes of non-denaturational character in DNA. This is in clear contrast with modification of DNA by clinically ineffective *trans*-DDP or monofunctional dienPt, both of which induce denaturational conformational alterations in DNA [69–71].

In this respect, *trans*-iminoether complexes resemble the antitumor *cis*-DDP which induces local conformational changes of non-denaturational character in DNA.

These results were confirmed fully by differential pulse polarography analysis, which distinguishes readily and with a great sensitivity between non-denaturational and denaturational conformational alterations induced in DNA by various physical or chemical agents. At a relatively low level of platination (r_b <0.05) *trans*-iminoether complexes, like *cis*-DDP and other antitumor analogs, induce non-denaturational conformational distortions in DNA; in contrast, *trans*-DDP, monofunctional dien-Pt, and other inactive platinum(II) complexes induce denaturational conformational alterations in DNA [69, 72].

4.5.3
Influence on the B→Z transition of DNA

The *B*→*Z* transition of DNA modified by platinum(II) complexes has also attracted considerable interest, because of a possible relationship with the molecular mechanism of the antitumor activity of these metal-based compounds. It has been shown that *cis*-DDP somewhat facilitates the *B*→*Z* transition induced by increasing NaCl concentration and radically lowers its cooperativity [73–78]. *trans*-Iminoether complexes were found to affect the *B*→*Z* transition only slightly [79]. This behavior was fundamentally different from that of clinically ineffective *trans*-DDP, which hinders the *B*→*Z* transition and lowers its cooperativity, as well as from that of dienPt, which stabilizes the *Z* form (observed at lower salt concentration than in the non-treated polymer) and radically reduces the transition cooperativity [73, 80].

4.5.4
Structure and Reactivity of Adducts in Specifically Modified Single- and Double-Strand Oligonucleotides

The chemical reactivity of monofunctional adducts was investigated in single- and double-strand oligonucleotides specifically modified by *trans*-iminoether complexes. Monofunctional adducts in single-strand oligonucleotides rearrange to several distinct bifunctional adducts in a rather slow reaction. In contrast, monofunctional adducts are stable in double-strand DNA and do not form inter-strand cross-links, clearly confirming the stability of monofunctional adducts of *trans*-iminoether complexes in double-strand DNA. Moreover the distortions induced by monofunctional adducts in double helical oligonucleotides are selectively located at the 5'-site with respect to the adduct, and involve the two complementary residues within the adjacent base-pair (Scheme 11).

4.6
Monofunctional Adducts Appear to Inhibit Transcription

DNA, globally modified by *trans*-iminoether complexes, terminates RNA synthesis prematurely with an efficiency similar to DNA adducts of *cis*-DDP. The prevalent lesions formed on DNA by *trans*-iminoether complexes are monofunctional adducts [59, 60] and it has been shown that monofunctional DNA adducts of *cis*-DDP, *trans*-DDP, and dienPt do not terminate RNA synthesis [81, 66, 71]. Thus the monofunctional adducts of *trans*-iminoether complexes on the one hand, and those of dienPt, *cis*-DDP, and *trans*-DDP on the other hand, modify the conformation of DNA in a fundamentally different manner.

5
The Impact of Platinum-Drugs on Cell Machinery

5.1
DNA-Adduct Formation

The biochemical mechanism by which *cis*-DDP determines cell death is not fully understood. It is generally believed that the antitumor activity of the drug is re-

^{5}T T C T C T TCTAG T C T TCT CTC ^{5}C T C T T C T C T T G T T C TC CTCT

^{3}A A G AGAAGATC AG A AGAGAG ^{3}GA G A A G AG A AC A AG AGGAGA

Filled symbols indicate strong reactivity of the corresponding base with DEPC (A) and with OsO_4 (T). G indicates platinated base.

Scheme 11

lated to structural and functional modifications of DNA stemming from the most prevalent 1,2-d(GpG) and 1,2-d(ApG) intra-strand cross-links, although those stemming from the inter-strand cross-links cannot be excluded [61, 62]. trans-DDP is stereochemically incapable of forming 1,2-d(GpG) and 1,2-d(ApG) adducts [61], suggesting that the difference in antitumor activity may result from the different nature of distortions induced in DNA by the various intra-strand cross-links. Also different are the inter-strand cross-links formed by cis-DDP (between guanine bases [81]), and by trans-DDP (between complementary guanine and cytosine residues [66]) which can therefore have a different biological effect [82, 83]. It has also been suggested that inactivity of trans-DDP may result from a high proportion of monofunctional adducts which may rapidly react with glutathione before they can be converted to more toxic bifunctional adducts [66, 84], and/or from preferential recognition repair of trans-DDP induced DNA adducts [85].

It has been suggested that reduced capacity to repair cis-DDP-damaged DNA contributes to the hypersensitivity of testis tumor cells to DNA-damaging agents [86].

5.2
Inhibition of Transcription

A two- or three-fold higher level of transcription was observed in human or hamster cell lines from plasmids containing trans-DDP adducts than from plasmids modified by cis-DDP. More efficient excision repair of the trans-DDP adducts was not the cause of its lower ability to block transcription; instead, trans-DDP adducts appeared to be preferentially bypassed by RNA polymerases: four times as many trans-DDP adducts than cis-DDP adducts were required to inhibit transcription elongation by 63% [87].

The results of such a study show that each of the predominant intra-strand adducts formed by cis-DDP, including 1,2-d(GpG), 1,2-d(ApG), and 1,3-d(GpTpG) cross-links – as well as inter-strand cross-links at d(GC) sites – provides an absolute block to RNA polymerases when present on the transcribed strand. In contrast, the 1,3-d(GpNpG) adduct (ca. 40% of the total adduct formed) and the bulk of the remaining intra-strand adducts formed by trans-DDP, are bypassed in vivo. The trans-DDP inter-strand cross-links (ca. 20% of the total in vitro adduct spectrum) which block transcription elongation effectively in vitro may be responsible for a significant portion of the ca. 30% blocking efficiency found in vivo [87].

RNA synthesis is more critical for rapidly dividing tumor cells than for stationary cells, and the former type of tumor cells have been found to be more sensitive to cis-DDP treatment. If the inhibition of RNA synthesis is involved in cell death, trans-DDP would be less effective than cis-DDP [87].

5.3
Cell Death by Apoptosis

In many cases cells treated with cytotoxic levels of *cis*-DDP display the biochemical and morphological features associated with programmed cell death (apoptosis) [88–91]. The dying cells eventually exhibit loss of membrane integrity and fragmented DNA, and – on electrophoretic separation – the nucleosome ladders characteristic of apoptosis. These observations suggest that platinum-DNA damage may exert its cytotoxicity through an intra-cellular signaling pathway, rather than – as originally believed – through the disruption of DNA replication [92].

Both *cis* and *trans* ammine/amine platinum(IV) species have also been demonstrated to initiate apoptosis in CH1 human ovarian carcinoma cell line at physiologically relevant concentrations (2 h exposure to IC_{50}). The main cell-cycle effect was a slowdown in S-phase traverse, during which most of the apoptosis appeared to occur. Failure to overcome the G(2) block was another cause of death [93, 94].

5.4
Comparison Among Cisplatin, Transplatin, and *trans*-Iminoether Complexes

trans-DDP is also able to induce programmed death in cells, although a much greater concentration than that of *cis*-DDP is required. Therefore, it is of interest to compare the fate of cells exposed to equitoxic doses of *cis*-DDP, *trans*-DDP, and *trans*-EE. The results are shown in Fig. 2 [95].

At equitoxic doses (1 h exposure of P388 leukemia cells to 12, 60, and 10 µM concentration of *cis*-DDP, *trans*-DDP, and *trans*-EE, respectively, followed by 48 h of postincubation), cell viability decreases smoothly from the start in the

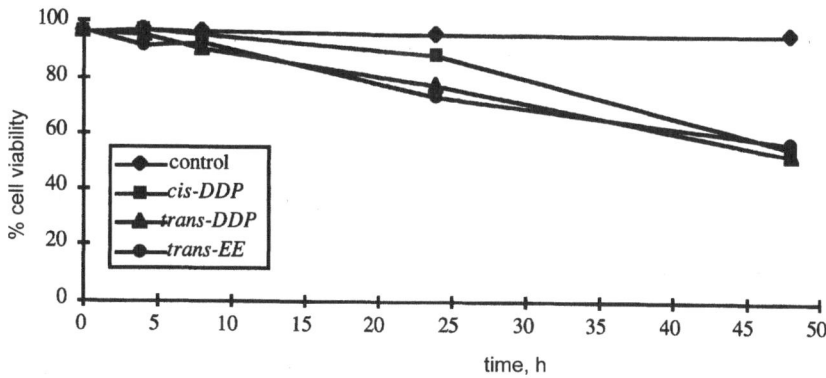

Fig. 2. Viability of P388 cells treated for 1 h with IC_{90} concentrations of *cis*- and *trans*-DDP and *trans*-EE. The percentage of viable cells was evaluated by the tripan-blue exclusion test at different post-treatment time intervals

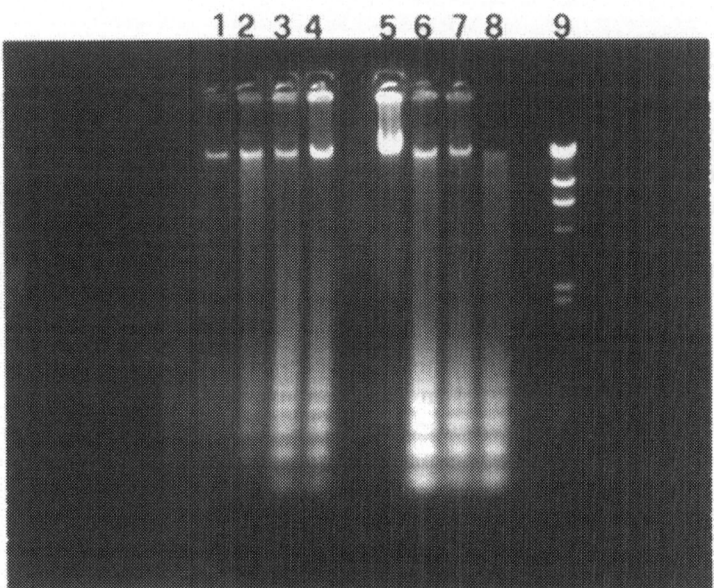

Fig. 3. Electrophoretic analysis of DNA isolated from P388 cells 24 h (*lanes 1–4*) and 48 h (*lanes 5–8*) after treatment for 1 h with IC_{90} of *cis*-DDP (*lanes 2* and *6*), *trans*-EE (*lanes 3* and *7*), and *trans*-DDP (*lanes 4* and *8*). DNA of control P388 cells (*lanes 1* and *5*) and of Hind III digest of λ phage as molecular weight marker (*lane 9*) are also included

case of *trans*-DDP and *trans*-EE (time dependence nearly identical for the two), while in the case of *cis*-DDP it decreases significantly only after 24 h postincubation.

Independently of the time of cell death, all treated cells die through an apoptotic mechanism, as shown by gel electrophoresis of DNA extracted from cells (Fig. 3).

Therefore, the mechanism of cell death appears to be similar in the three cases (apoptosis) and the time dependence is nearly identical for *trans*-DDP and *trans*-EE and indicates a faster response to drug action. Immediate inhibition of DNA synthesis at the earliest time point measured was also observed for *trans*-$[PtCl_2(py)_2]$; in contrast a longer incubation time was required for the *cis* isomer [27]. A linear time dependence of drug action may be an intrinsic characteristic of *trans*-platinum drugs in contrast to a delayed effect of the *cis*-analogs.

5.5
Analogies and Differences Among *trans*-Platinum Complexes

Although the effect of *trans*-DDP and *trans*-EE is comparable at equitoxic doses, a much greater concentration of *trans*-DDP is required in order to reach the same cytotoxicity as *trans-EE*. Pharmacokinetics and drug metabolism on one hand, and selective DNA interaction ultimately capable of triggering cell apoptosis on the other hand are the putative causes of the differences.

5.5.1
Stability in Biological Media

The inactivity of *trans*-DDP may relate to its chemical instability, that precludes delivery of the active compound to the tumor site, particularly in vivo.

Several observations in support of this hypothesis are as follows:

(i)   The additional stability stemming from the presence of two hydroxo ligands in the dichlorodihydroxoplatinum(IV) species appears to be sufficient to confer in vivo antitumor activity on *trans*-platinum complexes [24, 41], whereas their more reactive tetrachloro-platinum(IV) counterparts are inactive [41]. It has been shown, at least in the case of the two *cis* complexes tetraplatin and iproplatin (Scheme 3), that the tetrachloro species is rapidly and completely biotransformed to the platinum(II) reduction product within a few minutes in the plasma of drug-treated rats [96] while the dichlorodihydroxo species is reduced much more slowly [97].

(ii) Steric hindrance of the planar nitrogen heterocycles in *trans*-[PtCl$_2$(py)$_2$] and analogous complexes is expected to retard substitution kinetics relative to the NH$_3$ counterpart and may contribute to the enhanced cytotoxicity of these species [27].

(iii) Steric hindrance is also a characteristic of iminoether ligands in the third class of *trans*-active platinum compounds; it has been demonstrated that the latter compounds react more slowly with GSH and DNA than corresponding amine complexes [32].

It is noteworthy that sterically demanding carrier ligands have also been used to potentiate the antitumor activity of cis-oriented platinum(II) species. This is the case for *cis*-[PtCl$_2$(NH$_3$)(2-methylpyridine)] (AMD473) which has been selected for clinical trials in view of its reduced reactivity with sulfur ligands, its unique DNA binding properties, and its ability to circumvent several of the major resistance mechanisms manifested by cis-DDP resistant cell lines [98, 99].

5.5.2
Overcoming Resistance

A common feature of the *trans*-pyridine and *trans*-iminoether complexes was the greater intra-cellular platinum uptake. *trans* Species accumulate more rapidly and to a greater extent than their *cis*-counterparts and the DDP complexes [27]; in one case an immediate and irreversible drug-cell association prior to uptake has been hypothesized [27]. Altered drug uptake has been observed in *cis*-DDP-resistant murine cell lines [100, 101], as well as in human tumor lines of diverse origin [102, 103]. It follows that platinum complexes of novel structure may have an altered pharmacology of drug uptake and may be impervious to *cis*-DDP resistance mediated at the membrane level.

A role has also been suggested for thiols in platinum resistance. Elevated levels of GSH have been found in *cis*-DDP resistant cells [104, 105]. Likewise, enhanced metallothionein expression has been linked to *cis*-DDP resistance [106]. Glutathione may affect platinum complex cytotoxicity in one or two general ways:

(i)   by sequestering the complex prior to reaction with its biological target (DNA) and/or

(ii) by quenching and removing platinum-DNA monoadducts, thereby inhibiting formation of bidentate (toxic) lesions [84, 107].

A preliminary survey of the reaction of *trans*-pyridine and *trans*-iminoether complexes shows that the reaction is apparently slower than with *trans*-DDP. Also the *trans*-$[PtCl_2(OH)_2(amine)_2]$ complexes overcome acquired *cis*-DDP resistance, where resistance is mainly due either to reduced drug uptake or to enhanced platinum-DNA adduct removal [41]; this effect may be related to their stability in tissue-culture growth medium [41].

5.5.3
Interaction with DNA

The binding of the new antitumor active *trans*-platinum complexes to DNA has been examined in detail and placed in the context of those which had previously been elucidated for the DDP isomers. The main features of *trans*-pyridine complexes were: a great propensity to form inter-strand cross-links (a frequency of *ca.* 18% compared to <5% for the cisisomers) and the ability to cause a large unwinding angle (17° as compared with 13° for *cis*-DDP and 9° for *trans*-DDP). Both features may lead to differential repair in biological systems and to antitumor activity.

Unlike *trans*-DDP, the platinum(IV) compound JM335 (Scheme 6) [23] was efficient in forming inter-strand cross-links in SKOV-3 cells, when expressed either as absolute values or relative to the platination level of DNA. In contrast, inter-strand cross-links were not observed in CH1 cells. In the latter cells, single-strand breaks were observed after exposure to JM335. These unique DNA-binding properties of JM335 could account for its cytotoxicity and for its ability to circumvent acquired *cis*-DDP resistance resulting from increased repair/tolerance of DNA adducts.

The *trans*-iminoether complexes are quite distinct from the preceding two classes of complexes. They form stable monofunctional adducts at guanine bases which, differently from monofunctional adducts of DDP or dienPt, are able to interrupt RNA polymerases. These adducts do not appear to cause great conformational changes (small DNA unwinding and a very small effect on the midpoint and cooperativity of the $B \rightarrow Z$ transition induced by NaCl on double strand poly(dG-dC)), and are not recognized by antibodies specifically raised against DNA that had been modified by either *cis*- or *trans*-DDP [54].

Therefore, these studies call for a reevaluation of the structure-pharmacological activity relationships of platinum complexes which have been in general use until recently in the search for new platinum cytostatics. It has been demonstrated that *trans*-platinum drugs, which bind to DNA in a manner fundamentally different from that of *cis*-DDP can exhibit distinct – perhaps more efficient – anticancer activity than "conventional" platinum drugs.

6
Perspectives

In general, *trans*-DDP is less toxic to tumor cells than *cis*-DDP [108] by a factor of 5–20, and this is the main reason why *trans*-DDP is not useful as an antitumor agent. However certain carrier ligands – such as aromatic amines and iminoethers – in the *trans* configuration, may render platinum more active, and a series of analogs has been developed [25–27, 30–32]. Several *trans*-platinum(IV) species, which are metabolized to *trans*-platinum(II) compounds, are endowed with promising preclinical antitumor activity [23, 24]. For all these analogs, it may be expected that varying characteristics of cellular uptake, interaction with cytoplasmic nucleophiles, DNA-adduct formation, topological alterations of DNA, and interaction with DNA-repair systems will determine their cellular pharmacology. Understanding these characteristics will determine their optimal clinical use and will permit a broader application of platinum drugs to refractory tumors.

References

1. Rosenberg B, Van Camp L, Krigas T (1965) Nature 205:698
2. Greenwood NN, Earnshaw A (1986) Nickel, palladium and platinum. In Chemistry of the elements, Pergamon press, Oxford, p 1290
3. Rosenberg B, Van Camp L, Grimley EB, Thomson AJ (1996) J Biol Chem 242:1347
4. Loehrer PJ, Einhorn LH (1984) Ann Intern Med 100:704
5. Bosl GJ, Gluckman R, Geller NL, Golbey RB, Whitmore WFJr, Herr H, Sogani P, Morse M, Martini N, Bains M et al. (1986) J Clin Oncol 4:1493
6. Ozols RF (1995) Semin Oncol 22:61
7. Harrap KR (1995) Cancer Res 55:2761
8. Harrap KR (1985) Cancer Treat Rev 12:21
9. Calvert AH, Newell DR, Gumbrell LA, O' Reilly S, Burnell M, Boxall FE, Siddik ZH, Judson IR, Gore ME, Wiltshaw E (1989) J Clin Oncol 7:1748
10. Extra JM, Espic M, Calvo F, Ferme C, Mignot L, Marty M (1990) Cancer Chemother Pharmacol 25:299
11. Tashiro R, Kawada Y, Sakuri Y (1989) Biomed Pharmacother 43:251
12. Mathe G, Kidani Y, Sekiguchi M (1989) Biomed Pharmacother 43:237
13. Jennerwein MM, Eastman A, Khokhar A (1989) Chem Biol Interact 70:39
14. Saris CP, van de Vaart PJM, Rietbroek RC, Blommaert FA (1996) Carcinogenesis 17:2763
15. Levi F, Perpoint B, Garufi C, Focan C, Chollet P, Depres-Brummer P, Zidani R, Brienza S, Itzhaki M, Iacobelli S et al. (1993) Eur J Cancer 29 A:1280
16. Levy FA, Zidani R, Van Netzel JM (1994) J Natl Cancer Inst 86:1609

17. Diaz Rubio E, Sastre J, Zaniboni A, Labianca R, Cortes Funes H, de Braud F, Boni C, Benavides M, Dallavalle G, Homerin M (1998) Ann Oncol 9:105
18. Harder HC, Rosenberg B (1970) Int J Cancer 6:207
19. Pinto AL, Lippard SJ (1985) Biochim Biophys Acta 780:167
20. Plooy ACM, Fichtinger-Schepman AMJ, Schutte HH, van Dijk M, Lohman PHM (1985) Carcinogenesis 6:561
21. Fichtinger-Schepman AMJ, van der Veer JL, den Hartog JHJ, Lohman PHM, Reedijk J (1985) Biochemistry 24:707
22. Sherman SE, Lippard SJ (1987) Chem Rev 87:1153
23. Kelland LR, Barnard CFJ, Mellish KJ, Jones M, Goddard PM, Valenti M, Bryant A, Murrer BA, Harrap KR (1994) Cancer Res 54:5618
24. Mellish KJ, Barnard CFJ, Murrer BA, Kelland LR (1995) Int J Cancer 62:717
25. Farrell N, Ha TTB, Souchard J-P, Wimmer FL, Cros S, Johnson NP (1989) J Med Chem 32:2240
26. Van Beusichem M, Farrell N (1992) Inorg Chem 31:634
27. Farrell N, Kelland LR, Roberts JD, Van Beusichem M (1992) Cancer Res 52:5065
28. Zou Y, Van Houten B, Farrell N (1993) Biochemistry 32:9632
29. Farrell N (1996) Current status of structure-activity relationships of platinum anticancer drugs: activation of the *trans* geometry. In: Metal ions in biological systems, vol. 32, Marcel Dekker, New York, p 603
30. Coluccia M, Nassi A, Loseto F, Boccarelli A, Mariggiò MA, Giordano D, Intini FP, Caputo PA, Natile G (1993) J Med Chem 36:510
31. Coluccia M, Boccarelli A, Mariggiò MA, Caputo PA, Intini FP, Natile G (1995) Chem Biol Inter 98:251
32. Coluccia M, Mariggiò MA, Boccarelli A, Loseto F, Cardellicchio N, Caputo PA, Intini FP, Pacifico C, Natile G (1996) Iminoethers as carrier ligands: a novel *trans*-platinum complex possessing in vitro and in vivo antitumor activity. In: Pinedo HM, Schornagel JH (eds) Platinum and other metal coordination compounds in cancer chemotherapy, vol. 2. Plenum Press, New York, p 27
33. Farrell N (1996) DNA binding of dinuclear platinum complexes. In: Hurley LH, Chaires JB (eds) Advances in DNA sequence specific agents, vol 2. JAI Press, p 187
34. Farrell N (1995) Comments Inorg Chem 16:373
35. O'Dwyer PJ, Hudes GR, Walckzak J, Schilder R, La Creta F, Rogers B, Cohen I, Kowal C, Whitfield L, Boyd RA (1992) Cancer Res 52:6746
36. Kelland LR, Murrer BA, Abel G, Giandomenico CM, Mistry P, Harrap KR (1992) Cancer Res 52:822
37. Kelland LR, Abel G, McKeage MJ (1994) Proc Am Assoc Cancer Res 35:434
38. Mc Keage MJ, Mistry P, Ward J, Boxall FE, Loh S, O'Neill C, Ellis P, Kelland LR, Morgan SE, Murrer B et al. (1995) Cancer Chemother Pharmacol 36:451
39. Giandomenico CM, Abrams MJ, Murrer BA, Vollano JF, Barnard CFJ, Harrap KR, Goddard PM, Kelland LR, Morgan SE (1992) In Proceedings of the sixth international symposium on platinum and other coordination Compounds in cancer therapy, Plenum Press, San Diego, p 93
40. Hartwig JF, Lippard SJ (1992) J Am Chem Soc 114:5646
41. Kelland LR, Barnard CFJ, Evans IG, Murrer BA, Theobald BRC, Wyer SB, Goddard PM, Jones M, Valenti M, Bryant A, Rogers PM, Harrap KR (1995) J Med Chem 38:3016
42. Peyrone M (1846) Ann Chem Pharm 51:1
43. Kauffman GB, Cowan DO (1963) Inorg Synth 7:239
44. Barnard CFJ, Hydes PC, Griffith WP, Mills OS (1983) J Chem Res (S) 302
45. Basolo F, Bailar JC, Tarr BR (1950) J Am Chem Soc 72:2433
46. Kuroda R, Neidle S, Ismail IM, Sadler PJ (1983) Inorg Chem 22:3620
47. Giandomenico CM, Abrams MJ, Murrer BA Vollano JF, Rheinheimer MI, Wyer SB, Bossard GE, Higgins JD (1995) Inorg Chem 35:1015

48. Mellish KJ, Kelland LR, Barnard CFJ, Murrer BA, Harrap KR (1995) Proc Amer Ass Cancer Res 36:A2372
49. Novakova O, Vrana O, Kiseleva VJ, Brabec V (1995) Eur J Biochem 228:616
50. Goddard PM, Orr RM, Valenti MR, Barnard CFJ, Murrer BA, Kelland LR, Harrap KR (1996) Anticancer Res 16:33
51. Graves BJ, Hodgson DJ, Kralingen CG, Reedijk J (1978) Inorg Chem 17:3007
52. Cavallo L, Cini R, Kobe J, Marzilli LG, Natile G (1991) J Chem Soc Dalton Trans 1867
53. Paull KD, Shoemaker RH, Hodes L, Monks A, Scudeiro DA, Rubinstein L, Plowman J, Boyd MR (1989) J Natl Cancer Inst 81:1088
54. Wehrli W, Staehelin M (1971) Bacteriol Rev 35:290
55. Farrell N, Qu Y, Feng L, Van Houten B (1990) Biochemistry 29, 9522
56. Robert JD, van Houten B, Qu Y, Farrell N (1989) Nucleic Acids Res 17:9719
57. Keck MV, Lippard SJ (1992) J Am Chem Soc 114:3386
58. Cini R, Caputo PA, Intini FP, Natile G (1995) Inorg Chem 34:1130
59. Boccarelli A, Coluccia M, Intini FP, Natile G, Locker D, Leng M (1998) Anti-cancer Drug Design, accepted.
60. Brabec V, Vrana O, Novakova O, Kleinwächter V, Intini FP, Coluccia M, Natile G (1996) Nucleic Acids Res 24:336
61. Comess KM, Lippard SJ (1993) Molecular aspects of platinum-DNA interactions. In: Neidle S, Waring M (eds) Molecular aspects of anticancer drug-DNA interactions, vol 1. Macmillan Press, London, p 134
62. Mc A'Nulty MM, Lippard SJ (1995) Consequences of HMG-domain protein binding to cisplatin-modified DNA. In Eckstein F, Lilley DMJ (eds) Nucleic Acids and Molecular Biology, vol 9. Springer, Berlin Heidelberg New York, p 264
63. Zaludova R, Zakovska A, Kasparkova J, Balcarova Z, Vrana O, Coluccia M, Natile G, Brabec V (1997) Molecular Pharm 52:354
64. Bellon SF, Coleman JH, Lippard SJ (1991) Biochemistry 30:8026
65. Cohen GL, Bauer WR, Barton JK, Lippard SJ (1979) Science 203:1014
66. Brabec V, Leng M (1993) Proc Natl Acad Sci USA 90:5345
67. Lippard SJ, Ushay HM, Merkel CM, Poirier MC (1983) Biochemistry 22:5165
68. Vrana O, Kiseleva VI, Poverenny AM, Brabec V (1992) Eur J Pharmacol 226:5
69. Vrana O, Brabec V, Kleinwächter V (1986) Anti-Cancer Drug Design 1:95
70. Brabec V, Reedijk J, Leng M, (1992) Biochemistry 31:12397
71. Brabec V, Boudny V, Balcarova Z (1994) Biochemistry 32:1316
72. Brabec V, Kleinwächter V, Butour J-L, Johnson N (1990) Biophys Chem 35:129
73. Malfoy B, Hartmann B, Leng M (1981) Nucleic Acids Res 9:5659
74. Malinge JM, Leng M (1984) The EMBO Journal 3:1273
75. Malinge JM, Ptak M, Leng M (1984) Nucleic Acids Res 12:5767
76. Peticolas WL, Thomas GA (1985) J Mol Struct 126:509
77. Rahmouni A, Malinge JM, Schwartz A, Leng M (1985) J Mol Struct Dyn 3:363
78. Perez-Martin JM, Requena JM, Craciunescu D, Lopez MC, Alonso C (1993) J Biol Chem 268:24774
79. Zaludova R, Natile G, Brabec V (1997) Anti-Cancer Drug Design 12:295
80. Ushay HM, Santella RM, Caradonna JP, Grunberger D, Lippard SJ (1982) Nucleic Acids Res 10:3573
81. Lemaire M-A, Schwartz A, Rahmouni AR, Leng M (1991) Proc Natl Acad Sci USA 88:1982
82. Zhen W, Link CJ, O'Connor PM, Reed E, Parker R, Howell SB, Bohr V (1992) Mol Cell Biol 12:3689
83. Johnson SW, Perez RP, Godwin AK, Yeung AT, Handel LM, Ozols RF, Hamilton TC (1994) Biochem Pharmacol 47:687
84. Eastman A, Barry MA (1987) Biochemistry 26:3303
85. Ciccarelli RB, Solomon MJ, Varshavsky A, Lippard SJ (1985) Biochemistry 24:7533

86. Koberle B, Grimaldi KA, Sunters A, Hartley JA, Kelland LR, Masters JRW (1997) Int J Cancer 70:551
87. Mello JA, Lippard SJ, Essigmann JM (1995) Biochemistry 34:14783
88. Barry MA, Behnke CA, Eastman A (1990) Biochem Pharmacol 40:2353
89. Eastman A (1996) The interaction of cisplatin with signal transduction pathways and the regulation of apoptosis. In: Pinedo HM, Schornagel JH (eds) Platinum and Other Metal Coordination Compounds in Cancer Chemotherapy, vol 2. Plenum Press, New York, p 27
90. Sanchez-Perez I, Murguia JR, Perona R (1998) Oncogene 16:533
91. Siddik ZH, Mims B, Lozano G, Thai G (1988) Cancer Res 58:698
92. Demarq C, Bunch RT, Creswell D, Eastman A (1994) Cell Growth Differ 5:983
93. O'Neill CF, Ormerod MG, Robertson D, Titley JC, Cumber Walsweer Y, Kelland LR (1996) British J Cancer 74:1037
94. Ormerod MG, Orr RM, O'Neill CF, Chwalinski T, Titley JC, Kelland LR, Harrap KR (1996) British J Cancer 74:1935
95. Boccarelli A, Coluccia M, Intini FP, Natile G, unpublished results
96. Carafagna PF, Poma A, Wyrick SD, Holbrook DJ, Chaney SG (1991) Cancer Chemother Pharmacol 27:335
97. Pendyala L, Krishnan BS, Walsh JR, Arakali AV, Cowens JW, Creaven PJ (1989) Cancer Chemother Pharmacol 25:10
98. Raynaud FI, Boxall FE, Goddard PM, Valenti M, Jones M, Murrer BA, Abrams M, Kelland LR (1997) Clin Cancer Res 3:2063
99. Holford J, Raynaud FI, Murrer BA, Grimaldi K, Hartley JA, Abrams M, Kelland LR (1998) Anticancer Drug Design 13:1
100. Waud WR (1987) Cancer Res 47:6549
101. Kraker AJ, Moore CW (1988) Cancer Res 48:9
102. Andrews PA, Velury S, Mann SC, Howell SB (1988) Cancer Res 48:68
103. Bungo M, Fujiwara Y, Kasahara K, Nakagawa K, Ohe Y, Sasaki Y, Irino S, Saijo N (1990) Cancer Res 50:2549
104. Hospers GAP, Mulder NH, de Jong B, de Ley L, Uges DRA, Fichtinger-Schepman AMJ, Scheper RJ, de Vries EGE (1988) Cancer Res 48:6803
105. Teicher BA, Holden SA, Herman TS, Alvarez Sotomayer E, Khandekar V, Rosbe KW, Brann TW, Korbut TT, Frey E (1991) Int J Cancer 47:252
106. Kelley SL, Basu A, Teicher BA, Hacker MP, Hamer DH, Lazo JS (1988) Science 241:1813
107. Bancroft DP, Lepre CA, Lippard SJ (1990) J Am Chem Soc 112:6860
108. Macquet JP, Butour JL (1983) J Natl Cancer Inst 70:899

Chemistry and Biology of Multifunctional DNA Binding Agents

Nicholas Farrell, Yun Qu and John D. Roberts

Departments of Chemistry and Medicine, Virginia Commonwealth University, Richmond, Virginia

Dinuclear and trinuclear platinum complexes represent a new class of anticancer agents, distinct in DNA binding and antitumor activity from their mononuclear counterparts. The first representative of this class has now advanced to Phase I clinical trials. Dinuclear compounds may be differentiated amongst themselves with respect to important parameters of biological activity. This review summarizes the chemistry and biology of multifunctional DNA-binding agents, principally those characterized by the presence of two cis-$[PtCl_2(amine)_2]$ moieties in the same molecule.

Keywords. Multifunctional platinum; DNA binding; antitumor agents

1
Introduction

Modern chemotherapy for the treatment of cancer began in the late 1940s with the first DNA alkylating agent, nitrogen mustard. An understanding of the biological mechanism of alkylating agents subsequently led to their systematic development. A number have entered into full clinical use, with the most recent being ifosfamide in 1990. Approximately a dozen approved drugs and their recommended uses are now listed in chemotherapy handbooks [1]. The advent of the simple inorganic compound, cisplatin (cis-$[PtCl_2(NH_3)_2]$, cis-DDP), has similarly spawned research into "second" and third-generation analogs. The World Health Organization recognizes cisplatin for the treatment of germ-cell cancers, gestational trophoblastic tumors, epithelial ovarian cancer and small cell lung cancer as well as for palliation of bladder, cervical, nasopharyngeal, oesophageal and head and neck cancers.

The understanding that cisplatin acts by delivering the cis-$[Pt(NH_3)_2]$ moiety to DNA provided a rationale for systematically altering the structure of the platinum coordination sphere to generate altered activity/toxicity profiles. While only carboplatin has entered full clinical use, a number of analogs such as oxaliplatin, ZDO473, JM216 and Lobaplatin are currently being evaluated in Phase I and Phase II clinical trials(Fig. 1). A number of other platinum drugs that previously entered clinical trials, such as Tetraplatin (Ormaplatin), Iproplatin (CHIP) and CI- 473, are no longer under development, because of unacceptable toxic side effects, insufficient efficacy or limited spectrum of anticancer activity. It is too early to say whether the diversity of alkylating agents can be repeated within the platinum family.

Platinum coordination compounds based on a dinuclear motif represent a distinct class of DNA-modifying anticancer agents, Fig. 2. The general structure is best characterized as two platinum coordination spheres linked by a flexible-chain diamine. Systematic variation of coordination sphere, chain length and steric effects within the linkers produces a wide range of possible structures for biological evaluation. The rationale for development has been outlined in detail in recent reviews [2,3]. Briefly, it was our hypothesis that all direct structural analogs of cisplatin produce a very similar array of adducts on DNA and therefore induce similar biological consequences. This latter consideration led us to challenge the empirical structure-activity relationships and propose that platinum compounds structurally dissimilar to cisplatin may form different types of Pt-DNA adducts and thereby generate a spectrum of clinical activity genuinely complementary to the parent drug. The first drug to emerge from this work commenced clinical trials in June 1998.

Adducts formed in DNA by dinuclear compounds are also structurally diverse, with the possibility ranging from bifunctional to tetrafunctional DNA-binding. A survey of the structurally different compounds represented in Fig. 2 indicated the so-called 1,1/t,t series as having the most unique profile of antitumor activity. Systematic modification within this series has led to the current

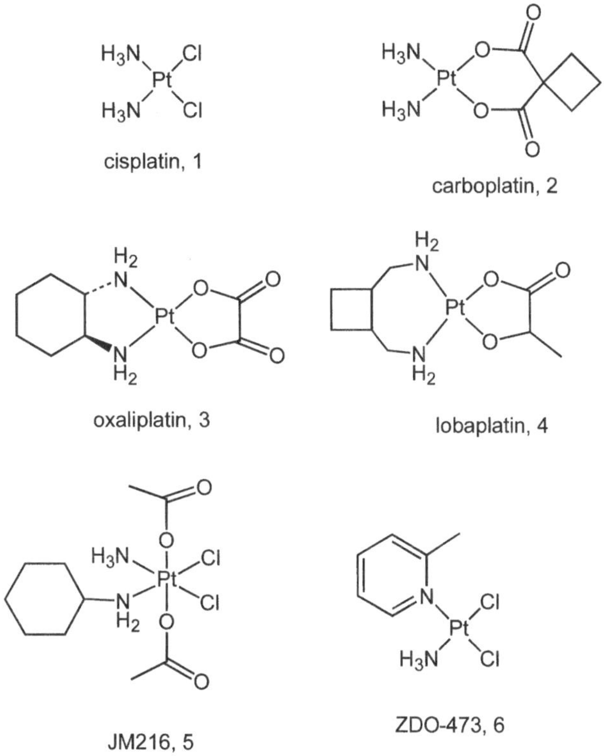

cisplatin, 1

carboplatin, 2

oxaliplatin, 3

lobaplatin, 4

JM216, 5

ZDO-473, 6

Fig. 1. Structures of clinically approved platinum coordination compounds (1 and 2) and representative compounds undergoing clinical trials (3-6)

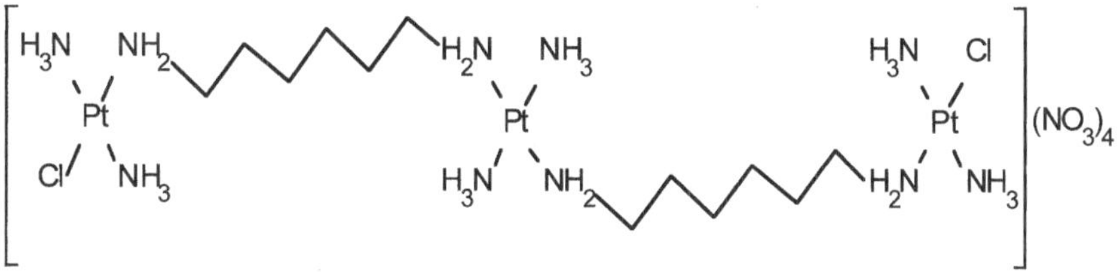

BBR3464

clinical development of the compound denoted BBR3464. Its structure is based on the dinuclear concept and is best described as two $trans$-$[PtCl(NH_3)_2]$ units linked by a tetra-amine $[Pt(NH_3)_2\{H_2 N(CH_2)_6NH_2\}_2]$:

This agent is the first genuinely new platinum agent not based on the "classical" cisplatin structure to enter clinical trials. The 4+ charge, the bifunctional

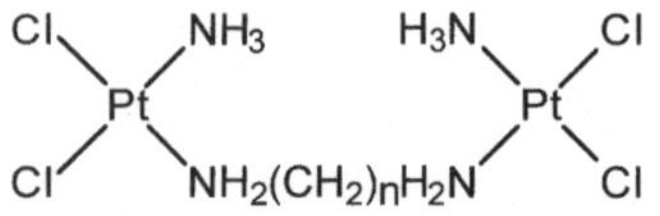

Fig. 2. Structures of the discrete classes of dinuclear compounds studied for DNA binding and antitumor activity. Abbreviations reflect number of substitution-labile chlorides on each platinum atom and their geometry with respect to the diamine bridge (See Ref 15 for details).

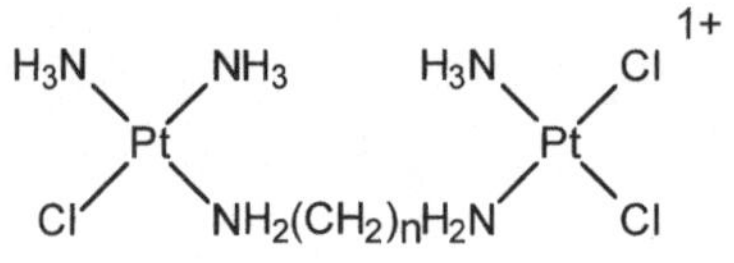

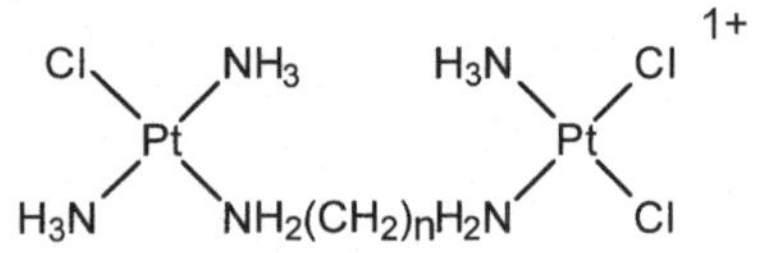

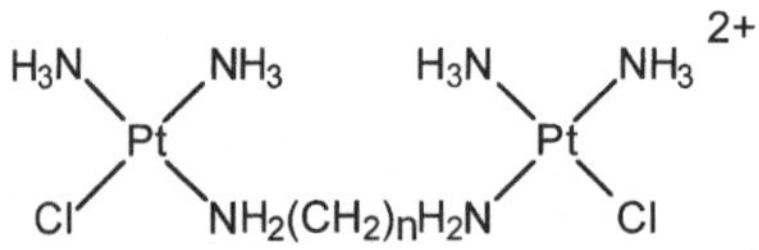

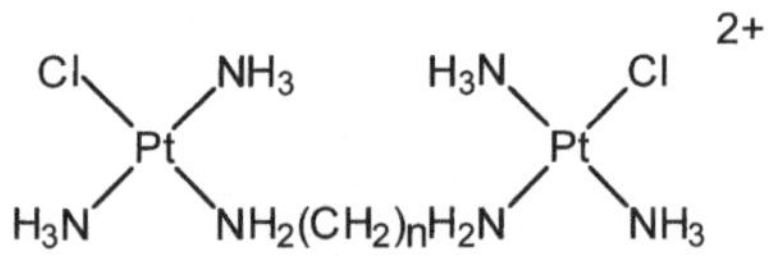

DNA binding over large distances, and the consequences of such DNA binding suggest an alteration in the paradigm of cisplatin-based antitumor agents. Designed synthesis of *dinuclear* platinum complexes with hydrogen-bonding ligands such as spermine (total charge 4+) and spermidine (total charge 3+) linkers mimics the essential biological features of BBR3464 [4]. The chemistry and biology of this new type of drug has been reviewed recently and will not be repeated in detail here [5,6]. Table 1 summarizes the highly promising antitumor activity.

It is, however, instructive to ask the question how the dinuclear structures, representing bifunctional, trifunctional and tetrafunctional DNA-binding agents, differ amongst themselves with respect to profiles of biological activity. Multifunctional DNA binding modes can be considered to consist of more than two possible covalent binding sites. A logical extension of our initial argument is that there may also exist differential biological activity within the various series represented in Fig. 2. The question to be addressed is whether all these structures are similar or if biological activity is affected within the discrete classes comprising the dinuclear structure. This review summarizes the data for multifunctional DNA-binding compounds.

Table 1. Comparision at Maximum Tolerated Dose of BBR 3464 (0.2-0.4 mg/kg) and Cisplatin (3-6 mg/kg) after i.v. repeated treatment on human solid tumors

Clinical Parameter[a]	BBR 3464	Cisplatin
Resistant TWI < 50%	0	9 (4 NSCLC, 2 ovarian, 2 gastric, 1 prostatic)
Responsive TWI 50-70%	3 (1 NSCLC, 1 gastric, 1 prostatic)	7 (2 SCLC, 2 NSCLC, 2 ovarian, 1 bladder)
Sensitivity, TWI > 70%	15 (3 SCLC, 5 NSCLC, 5 ovarian, 1 gastric,1 bladder)	2 (1 ovarian, 1 SCLC)

a TWI% is Tumor Weight Inhibition compared to controls.SCLC: small cell lung cancer. NSCLC: non-small cell lung cancer. The clinical parameter refers to the fact that clinical resistance, relative resistance and sensitivity are most likely to be seen at these TWI levels – a drug achieving a TWI% of < 50% is unlikely to have significant clinical efficacy in that tumor type. Thus, of 18 tumors tested, 15 appear sensitive to BBR3464 while 3 are responsive. In the same panel of 18 tumors, only two are truly sensitive to cisplatin. The sensitivity of tumors difficult to treat in the clinic such as gastric and bladder is greater to BBR3464 than to cisplatin. Note that the maximum tolerated dose of BBR 3464 is significantly less than that of cisplatin – approximately thirtyfold on a molar basis. Most direct cisplatin structural analogs (See Figure 1) need significantly higher doses to achieve similar potency to the parent drug.

2
Chemistry and DNA Binding of Tetrafunctional Dinuclear Platinum Compounds

2.1
Synthesis and Model Studies

Compounds of formula $[\{cis\text{-}PtCl_2(NH_3)\}_2(NH_2(CH_2)_nNH_2)]$, (2,2/c,c), are tetrafunctional. They are usually prepared by reaction of two equivalents of a suitable mononuclear platinum complex with the diamine:

$2 K[PtCl_3(NH_3)]+NH_2(CH_2)_nNH_2 \rightarrow [\{cis\text{-}PtCl_2(NH_3)\}_2(NH_2(CH_2)_nNH_2)]$
where (n>4)

Their lack of water solubility (a general problem for compounds containing PtN_2Cl_2 coordination spheres) precluded extensive development of this series (see below). Substitution of the chlorides with water-solubilizing groups such as malonate etc. is relatively straightforward [7]:

$[\{cis\text{-}PtCl_2(NH_3)\}_2(NH_2(CH_2)_nNH_2)]+2 Ag_2(mal) \rightarrow [\{Pt(mal)(NH_3)\}_2(NH_2(CH_2)_nNH_2)] +4AgCl$

These compounds, Fig. 3, offer the advantage of water solubility but present the interesting problem that DNA substitution is likely to occur preferentially on only one Pt center. The first DNA-binding step must involve breaking the chelate ring, exactly analogous to the situation for monomeric compounds such as carboplatin [8]. Model studies on monomeric systems indicate that displacement of monodentate carboxylate is significantly faster than the reaction to break the

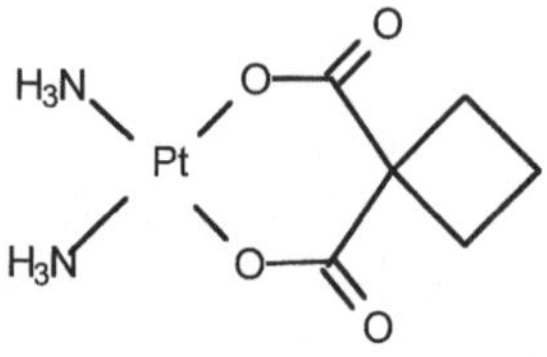

Carboplatin, 1 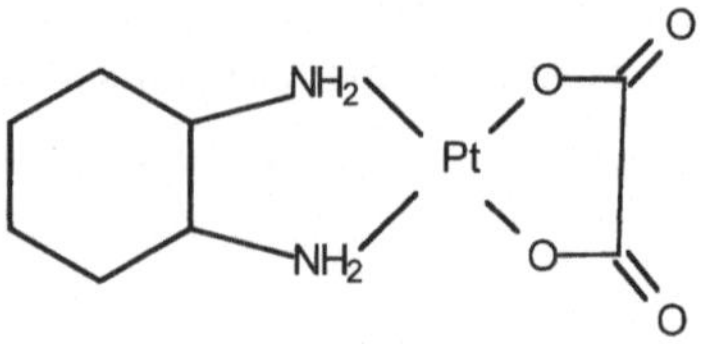   Oxaliplatin, 3

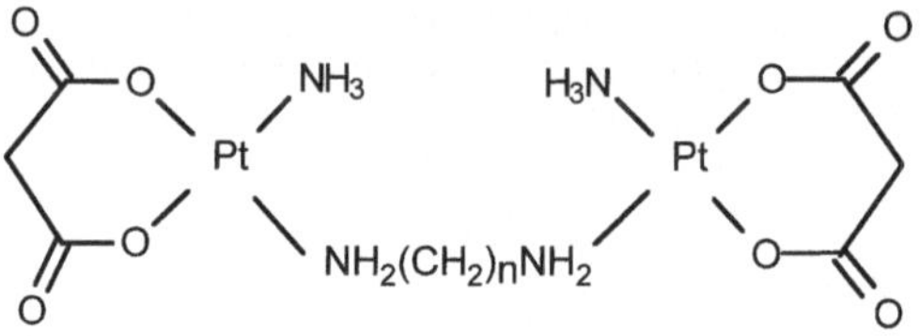

2,2/c,c(malonato), 12

Fig. 3. Structure of dinuclear malonate compounds (12) and their direct mononuclear analogs

chelate ring [9]. It might be expected, then, that the second DNA-binding step would be predominantly that of preferential displacement of a monofunctional carboxylate ligand. Evidence for this mode of DNA attack has been obtained from model studies of the interaction of 5'-GMP with the dinuclear malonate species. NMR chemical shifts indicative of the preferential formation of the $Pt(5'\text{-}GMP)_2$ coordination sphere were observed [10]:

$$[\{Pt(mal)(NH_3)\}(NH_2(CH_2)_nNH_2)\{Pt(mal)(NH_3)\} + 2\ 5'\text{-}GMP \rightarrow [\{Pt(5'\text{-}GMP)_2$$
$$(NH_3)\}(NH_2(CH_2)_nNH_2)\{Pt(mal)(NH_3)\}\ (\text{major product})$$

Two further, interesting classes of dinuclear platinum compounds containing *cis*-[PtCl$_2$(amine)$_2$] coordination spheres have also been synthesized by other groups, Fig. 4 [11,12]. In the case of the ethylenediamine-based species there is the potential for chiral effects on the substituted amine group forming the linker. Solvolysis in DMSO gave interesting mixtures of *cis/trans* isomers by substitution of the distinct Pt-Cl bonds [12].

2.2
DNA Binding

Upon cisplatin binding to DNA, monofunctional adducts and bifunctional adducts such as d(GG), d(AG) and d(GNG) intrastrand crosslinks are formed.

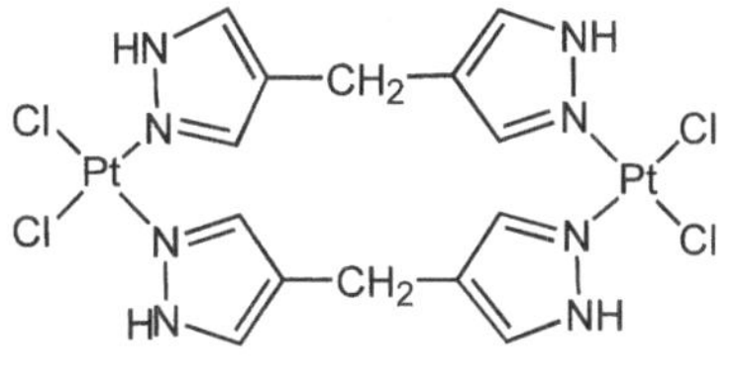

13

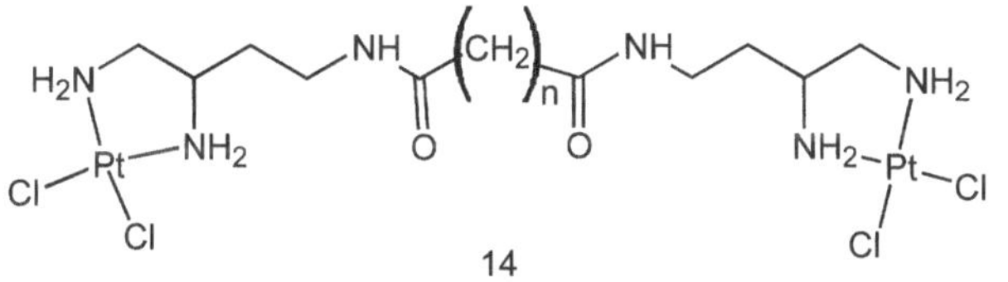

14

Fig. 4. Structure of dinuclear platinum compounds containing sterically rigid linkers (13) and based on the ethylenediamine chelate (14).

DNA-DNA interstrand GG cross-links and DNA-protein crosslinks are produced to a much lesser extent. Inhibition of DNA synthesis and transcription has been measured as consequences of Pt-DNA adduct formation. The individual adducts differ in their ability to inhibit specific processes essential for cell growth and division [13,14].

In contrast, the predominant DNA binding mode of dinuclear platinum compounds is DNA-DNA interstrand crosslinking where one Pt coordination sphere binds to one strand of DNA [15]. The extent of interstrand cross-linking is dependent on the exact structure of the compound. At r_b >0.025 the 1,1/t,t compound produces significantly more interstrand crosslinks than its 2,2/c,c counterpart. This result is in interesting contrast to overall binding where the 2,2/c,c compound increases linearly with increasing Pt concentration whereas the r_b for the 1,1/t,t compound rises to a maximum and then falls off [16]. These results are now consistent with the model studies above where the binding of 2,2/c,c compounds is best considered as a combination of cisplatin-like intrastrand and dinuclear (Pt,Pt) interstrand crosslinks. Sequencing studies also indicated distinct binding preferences of both dinuclear compounds in comparison to cisplatin but binding sites are not the same for both compounds.

DNA-DNA interstrand crosslinking requires in principle only one substitution-labile Pt-Cl bond on each Pt center, and the minimum requirement is thus bifunctionality. At this level therefore, there is no significant difference between the potentially bifunctional (1,1/t,t) and tetrafunctional (2,2/c,c) compounds. Two unique properties of tetrafunctional platinum compounds are their ability to (a) produce DNA-protein ternary crosslinks [17] and (b) interhelical DNA-DNA crosslinks, where two double helices are linked together [18].

2.3
Ternary DNA-Protein Cross-Link Formation

In the case of DNA-protein crosslinking, definitive proof was shown that a (Pt,Pt) interstrand crosslink acted as the precursor for ternary DNA-protein complex formation, as per the general case of Scheme 1.

For a dinuclear platinum compound $M_a=M_b=Pt$. To extend this chemistry the heterodinuclear compound, Fig. 5, was synthesized:

cis-[PtCl$_2$(NH$_3$){H$_2$N(CH$_2$)$_4$NH$_3$}]$^+$+cis-[RuCl$_2$(DMSO)$_4$] → [{cis[PtCl$_2$(NH$_3$)} H$_2$N(CH$_2$)$_4$NH$_2${ cis-RuCl$_2$(DMSO)$_3$].

The advantage of mixed-metal DNA-protein binding agents is that greater selectivity may be introduced. It is possible that the inertness of a Ru(II) or Ru(III) center relative to Pt(II) would allow incorporation of specific Pt-DNA binding modes to be followed in a second step by Ru-protein bond formation. Compound 15 had moderate antitumor activity with a T/C% of 152 @ 25 mg/kg in P388 leukemia. Its sensitivity to light and hydrolytic reactivity precluded the

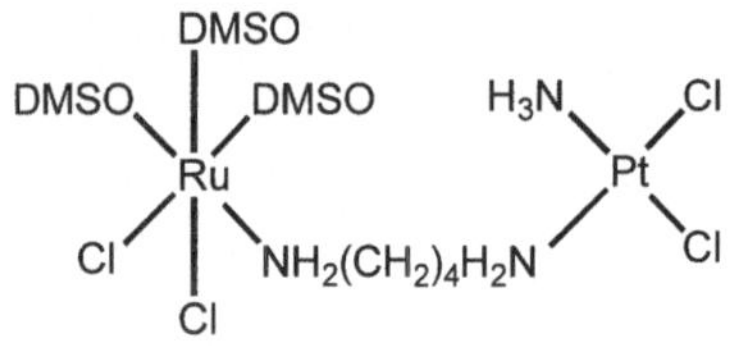

15

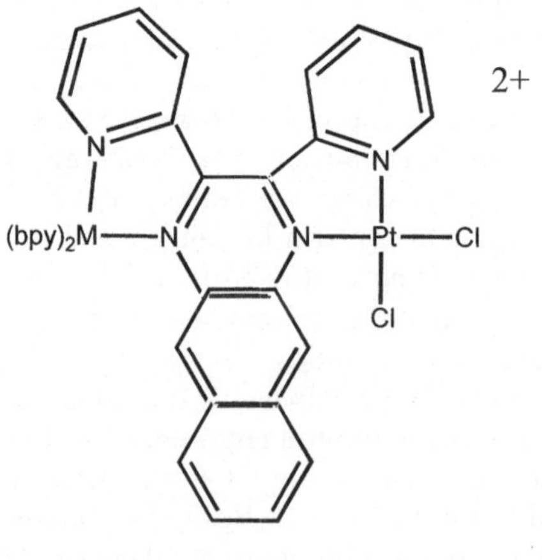

16

Fig 5. Structures of a heterodinuclear (Pt,Ru) DNA-protein crosslinking agent, (15), and heterodinuclear (M = Ru, Os) DNA binding agents (16) based on the M(bpy)2 (bpy - 2,2'-bipyridine) moiety.

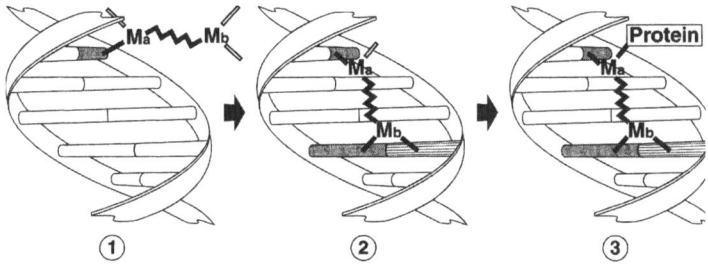

Scheme 1

further use of this compound as a drug or probe of DNA-protein interactions [19]. Measurement of intracellular uptake and intracellular DNA binding showed an almost three-fold excess of Pt over Ru [20]. These results indicate a rapid breakdown of the compound in tissue culture, in accord with the chemical studies. Other approaches to heterodinuclear (Pt,Ru) DNA-binding agents involve incorporation of the Ru center into more stable coordination environments such as the $Ru(bpy)_2$ moiety (M=Ru in **16**, Fig. 5). These compounds may have interesting properties due to coupling of the electron transfer and light absorbing properties of the Ru and Os centers with the DNA binding of the platinum [21,22].

Observation of Metal-DNA-protein ternary complex formation raises the possibility that such a unique adduct could act as a "suicide lesion" irreversibly sequestering, for example, a repair protein or transcription factor [17]. Cisplatin forms relatively few ternary DNA-protein crosslinks [23]. The effect of cisplatin-DNA binding upon protein recognition is most commonly viewed in terms of "hijacking" proteins or blocking access of proteins to specific binding sites [24]. *In situ*, cross-linking by cisplatin of nuclear matrix-bound transcription factors has recently been observed in nuclear DNA of human breast cancer cells [25]. These results suggest an attractive mechanism by which cisplatin inhibits transcription and replication processes. Sterically hindered amine ligands in the mononuclear *cis*-$[PtCl_2(amine)_2]$ geometry have also been used to enhance ternary DNA-protein crosslink formation relative to cisplatin [26]. Trifunctional dinuclear compounds (**8** and **9**, Fig. 2) are also very effective at DNA-protein crosslinking.

2.4
Interhelical DNA-DNA Cross-Linking

The possible formation of interhelical crosslinks was first mentioned in early examinations of DNA binding by tetrafunctional dinuclear platinum complexes [27]. A more definitive confirmation of this has recently been reported using a new set of dinuclear platinum complexes based on a coordinating ethylenediamine moiety, See Fig. 3. These compounds share many of the features elaborated for the compounds developed in our laboratory but also some new effects

INTERHELICAL CROSSLINKING

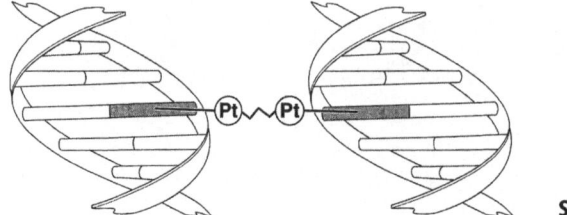

Scheme 2

were characterized. Analysis of products of interstrand crosslinking on denaturing polyacryalamide gels showed distinct areas due to interstrand crosslinking between dsDNA molecules – an interhelical crosslink, Scheme 2.

3
Biological Activity of Dinuclear Platinum Compounds

3.1
In Vitro Cytotoxicity in a Human Ovarian Panel

Tetrafunctional dinuclear platinum compounds showed good antitumor activity in murine leukemia L1210 and P388 models resistant to cisplatin (L1210/DDP and P388/DDP) [7]. Studies varying the chain length of the diamine linker indicated that an optimal value of n=4 to 8 (butanediamine to octanediamine) resulted in an equivalent antitumor effect. Comparison of malonate versus chloride as leaving ligands indicated that the potency of the malonates was lower but they showed better *in vivo* activity in selected tumor model systems such as murine colon 26 [7]. Interestingly, in cisplatin-resistant murine L1210 leukemia, potency and antitumor activity were decreased for the malonates relative to chlorides, suggesting a cell line/tumor type dependence for activity in the malonate compounds.

The biological activity of two representative (1,1/t,t and 2,2/c,c) dinuclear compounds against a panel of human ovarian xenografts is summarized in Table 2. Analysis of trends throughout the panel allows comparison of patterns of activity for different drugs [28]. It is apparent that activity profile is similar between cisplatin, carboplatin and the 2,2/c,c compound, but differences are apparent between these three compounds and the charged 1,1/t,t compound. This trend reflects the clinical situation where both cisplatin and carboplatin display similar efficacy in ovarian cancer. The results confirm that the two dinuclear compounds do indeed behave differently. The 2,2/c,c compound is also clearly more potent than its direct counterpart – carboplatin.

Table 2. Comparison of In Vitro Cytoxicity of Dinuclear and Mononuclear Platinum Compounds in a Human Ovarian Tumor Panel[a]

Compound	HX/62	SKOV-3	PXN/94	41M/CISR	41M	CH1/CISR	CH1
[{Pt(mal)(NH$_3$)}$_2$ NH$_2$(CH$_2$)$_4$NH$_2$]	44.5[a] ± 3.5	20.8 ± 3.7	9.9 ± 2.8	4.3 (3.1)[b] ± 1.4	1.4 ± 0.44	2.4 (4.4) ± 0.44	0.54 ± 0.03
[{PtCl(NH$_3$)$_2$}$_2$ NH$_2$(CH$_2$)$_4$NH$_2$]$^{2+}$	46 ± 14.7	57 ± 8.8	14.3 ± 1.8	5.1 (0.71) ± 0.78	7.2 ± 0.33	6.3 (2.4) ± 4.2	2.6 ± 1.4
cisplatin	12.6 ± 1.5	4.4 ± 1.4	3 ± 0.57	1.4 (6.1) ± 0.1	0.23 ± 0.03	0.67 (6.7) ± 0.1	0.1 ± 0.014
carboplatin	70 ± 7	38 ± 9	31 ± 5	10.4 (2.8) ± 0.9	3.7 ± 0.4	4.2 (4.2) ± 0.2	1 ± 0.2

a ID$_{50}$ (μM).
b RF is resistance factor defined as ratio of ID$_{50}$'s in sensitive and resistant cells. Data obtained as per Ref. 28.

3.2
Toxicity in Hypoxic and Aerobic Tissues

An interesting example of differences in biological activity within dinuclear compounds is demonstrated by their relative toxicity in hypoxic tissue. Tumors often have regions of hypoxia caused by poor vascularization. The absence of oxygen renders cells resistant to radiotherapy and, indeed, some forms of chemotherapy. Efforts to improve radiosensitization include use of "electron-affinic" nitroimidazoles as oxygen mimics [29]. In related work platinum complexes of nitroimidizoles have been prepared to target the "electron-affinic" organic moiety to the nucleus and DNA [30]. Recently, proof of this concept has been obtained by immunocytochemical labeling of aerobic and mammalian cells – in contrast to the free ligand which was localized in the cytoplasm, a Pt complex of the radiosensitizer EF5 was shown to be localized more in the nucleus [31].

Cisplatin and radiation is a useful combination in the clinic. Radiosensitization and toxicity of cisplatin is also greater in hypoxic than in aerobic cells [32]. Interestingly, the 2,2/c,c dinuclear compounds (as their malonates) also show selectivity in hypoxic cells [33]. The HCR (hypoxic cytotoxicity ratio) of approximately 3 is similar to that of cisplatin. This difference is not explained by enhanced accumulation or DNA binding alone. In contrast, bifunctional 1,1/t,t and 1,1/c,c compounds (See Fig. 2) do not display any hypoxic selectivity. The heterodinuclear (Pt,Ru) compound, **15**, displays similar hypoxic selectivity [20]. These results support the hypothesis of Skov that DNA-protein crosslinking may be an important determinant in hypoxic toxicity. The unique ability of the tetrafunctional dinuclear compounds to form DNA-protein ternary crosslinks may result in the biological consequences noted [32].

3.3
In Vivo Antitumor Activity

The enhanced potency of the dinuclear structure was also observed in human ovarian carcinoma IGROV-1 xenografts, Table 3 [38]. In this case, there appears to be little difference in chain length between the three dinuclear compounds studied – all are more potent than carboplatin at the equivalent dose. What is also noteworthy is the reduced toxicity of the dinuclear compounds relative to that of carboplatin, $[Pt(CBDCA)(NH_3)_2]$. No toxicity is observed at doses up to 80 mg/kg. Earlier studies indicated that safe doses up to 150 mg/kg were obtainable with the dinuclear compounds [7]. Nevertheless the potential of the 2,2/c,c malonate compounds to achieve similar efficacy in at least this xenograft is clear.

The malonate compound also showed significant activity in murine Colon26 tumor with 3/10 cures [7]. This remarkable result warranted confirmation in human colon tumor xenografts. At 60 mg/kg a TWI% of 53 was obtained, superior to cisplatin. The activity of the 2,2/c,c malonate in colon tumors is of interest due to the purported efficacy of oxaliplatin in combination with 5-FU in colon cancers – Phase II clinical trials of this combination are ongoing [34].

Table 3. Antitumor activity of neutral dinuclear platinum complexes, carboplatin and cisplatin against human ovarian IGROV-1 xenografts in nude mice (ip/ip,q4dx3)[a]

Compounds[b]	Dose (mg/kg/day)	T/C%[c]	Toxicity [d]
Cisplatin	3	183, 221, 238	0/25
	4	216 (128–289)	6/54
Carboplatin	60	172, 203	0/16
	80	119, 203	4/16
$[\{Pt(mal)(NH_3)\}_2H_2N(CH_2)_4NH_2]$	60	228	0/8
	80	250	0/8
$[\{Pt(mal)(NH_3)\}_2H_2N(CH_2)_5NH_2]$	60	247	0/8
	80	262, 289	0/15
$[\{Pt(mal)(NH_3)\}_2H_2N(CH_2)_6NH_2]$	60	228	0/8
	80	289	0/8

a CD1 nu/nu female mice were transplanted ip with 2.5×10^6 cells. Treatment was given ip on days 3,7,11 after tumor transplantation (day 0).
b All compounds were dissolved in sterile water.
c Median survival time of treated mice/median survival time of controls ×100.
d No. of toxic deaths/total No. of mice.

4
Cellular Pharmacology

The preceding sections have summarized results supporting the view that both the pattern of DNA binding and antitumor activity differ amongst the distinct structural classes of dinuclear compounds. The majority of DNA binding studies tend to be made firstly in non-cellular systems. Extrapolation to the "real" cellular system and proof of such extrapolation is essential both for confirmation of mechanistic hypotheses and for full understanding of a drugs potential. To examine this feature the cellular pharmacology of a representative tetrafunctional dinuclear compound was compared with cisplatin and carboplatin in L1210 cells sensitive and resistant to cisplatin (L1210/0 and L1210/DDP respectively). Cellular pharmacology parameters studied were cellular accumulation, Pt-DNA binding and interstrand crosslinking. The results are summarized in Table 4. In vitro cytotoxicity results confirmed previous reports. The dinuclear compound is somewhat more potent than its mononuclear analog and is more active also in the resistant line. Both dicarboxylate compounds are less potent than cisplatin, as expected.

4.1
Cellular Accumulation and DNA Binding

Altered cellular accumulation is a frequently described feature of cellular resistance to cisplatin. Accumulation generally was linear over the time points and concentrations studied. Accumulation of cisplatin is much greater than that of the dicarboxylate complexes in both cell lines. There is a significant decrease in accumulation of cisplatin in the resistant cell line. In contrast, accumulation of both carboplatin and the dinuclear compound was essentially unaffected by the resistance phenotype.

Table 4. Cellular pharmacology of a dinuclear malonate compound, caboplatin, and cisplatin in L1210 murine leukemia cells sensitive and resistant to cisplatin

	In Vitro Cytotoxicity ID$_{50}$ (μM)		Cellular Accumulation[b]		DNA Binding[c]	
Compound	L1210/0	RF[a]	L1210/0	L1210/DDP	L1210/0	L1210/DDP
[{Pt(mal)(NH$_3$)}$_2$NH$_2$(CH$_2$)$_4$NH$_2$]	3.8	1.4	96	108	363	288
carboplatin	7.2	7.7	134	134	342	83
cisplatin	0.42	27	770	256		

a RF is resistance factor calculated from ratio of ID$_{50}$'s in sensitive and resistant cells. See ref. 29 for details.
b fmol [Pt compound)/10^6 cells/μM [Pt compound} for 2 h exposure to cells.
c pmol [Ptcompound)/g DNA/μM [Ptcompound] for 2 h. The procedure followed that of Ref. 37.

A further determinant of cytotoxicity is the amount of platinum actually bound to intracellular DNA, which reflects both the inherent affinity of the drug for DNA as well as the effect of competing metabolic processes such as binding to plasma proteins and intracellular thiols, such as glutathione. Despite differences in molecular weights and cellular accumulation, DNA binding of the malonate compound is equivalent to that of carboplatin in L1210/0. Cellular accumulation of carboplatin is unaltered in L1210/DDP. Reduced DNA binding in this cell line is thus presumably related to a combination of enhanced DNA repair and enhanced intracellular sequestration by thiols such as glutathione. As DNA binding of the dinuclear compound is not significantly different in the two cell lines, these mechanisms of cellular resistance clearly do not have the same effect. The nature of the DNA adducts may be intrinsically more difficult to repair.

4.2
Interstrand Crosslinking in Cells

Does the observation of enhanced DNA-DNA interstrand crosslinking in laboratory experiments on plasmid DNAs reflect the situation inside cells? The technique of alkaline elution was used to probe this question in the L1210/0 cell line. Time course studies showed that the kinetics of interstrand crosslinking of the dinuclear compound was different to the mononuclear analogs, Fig. 6. Following a two-hour incubation with 2,2/c,c interstrand crosslinking of L1210 cellular DNA was greatest at the earliest time point observed and significant crosslinking persists at 26 hours. This contrasts with carboplatin for which interstrand crosslinking was minimal at 2 hours and peaked at 8–10 hours. These latter observations are in agreement with published work on cisplatin [35]. There is an approximate 5-fold difference between the maximal interstrand crosslinking of the dinuclear compound and carboplatin. The results indicate that enhanced cellular interstrand crosslinking is a real effect of the dinuclear structure. Further, their profile is significantly different to those formed by carboplatin.

5
Summary and Perspectives

Design and development of cisplatin analogs stressed the need for the *cis*-DDP structure, and varied both the nature of the leaving group (chloride) and non-leaving group (amine). These analogs, however, have not shown a greatly altered spectrum of clinical efficacy in comparison to the parent drug. The relative efficiency of the biological action of analogs in comparison to cisplatin (as indicated by assays measuring cell death or by monitoring DNA replication and transcription) is dependent on the kinetics of adduct formation as well as the specific structure of the Pt-DNA adduct formed.

The entry into the clinic of a totally new structural type such as BBR3464 and related dinuclear platinum compounds is an important breakthrough for platinum anti-cancer research. It is unlikely that the number of direct structural ana-

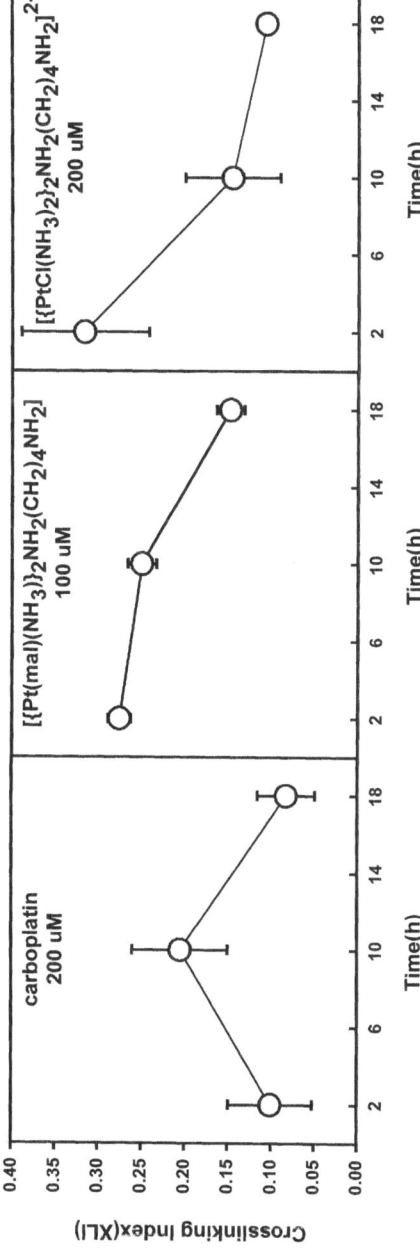

Fig. 6. Time course studies of alkaline elution of protease treated cellular DNA from L1210 luekemia cells following a 2h exposure to platinum compounds. Data points represent mean of 2 experiments. XLI indicates elution rate equal to control cells. Note that both dinuclear platinum compounds exhibit significantly greater interstrand crosslinking than carboplatin. Experimental protocol as per Ref. 35.

logs of cisplatin in the clinic will be increased. The results from the work on dinuclear compounds suggest that a clinically-useful family of platinum-based anticancer compounds should contain structurally-diverse members that are not directly analogous to cisplatin. In this respect, there is now strong evidence from a number of groups that the *trans*platinum geometry, hitherto considered inactive, may also afford interesting and active compounds [36]. Results cited within this review show that even within the family of dinuclear platinum compounds, there is differential biological activity. Multifunctional homo- and heterodinuclear compounds display consistently greater hypoxic cytotoxicity and are capable of unique DNA-protein interactions. Further research is warranted to delineate the clinical potential for exploitation of these differences.

Acknowledgments. It is a pleasure to acknowledge our many collaborators and coworkers as listed in the references. I thank Lloyd Kelland for data of Table 2 and Kirsten Skov and Hans Adomat for their collaboration. The group in Boehringer Mannheim Italia is especially acknowledged for the data of Table 3 and their continuing enthusiasm. This work is supported by grants from Boehringer Mannheim, National Science Foundation and The American Cancer Society.

References

1. Loeb S (Executive Director) (1994) Chemotherapy Handbook, Springhouse Corporation, Springhouse, PA.
2. Farrell N (1998) DNA Binding of Non-Classical Platinum Antitumor Complexes. In: Palumbo M (ed) Advances in DNA Sequence Specific Agents, Vol. 3. JAI Press Inc. New Haven p 179
3. Farrell N (1996) DNA Binding of Dinuclear Platinum Complexes. In: Hurley LH, Chaires JB (eds) Advances in DNA Sequence Specific Agents, vol 2. JAI Press Inc. p 187
4. Rauter H., Di Domenico R., Menta E, Oliva A, Qu Y, Farrell N (1997) Inorg Chem 36:3919
5. Farrell N, Qu Y, Bierbach U, Valsecchi M, Menta E (1999) Structure Activity Relationships in Bifunctional Dinuclear and Trinuclear Platinum Anticancer Agents In: Lippert B (ed) Cisplatin: Chemistry and Biochemistry of a Leading Anticancer Drug. Verlag Basel (in press)
6. Farrell N, Spinelli S (1998) Dinuclear and Trinuclear Platinum Anticancer Agents In: Farrell N (ed) Uses of Inorganic Chemistry in Medicine. Royal Society of Chemistry Cambridge England (in press)
7. Kraker AJ, Hoeschele JD, Elliott WL, Showalter HDH, Sercel AD, Farrell N (1992) J Med Chem 35:4526
8. Knox RJ, Friedlos F, Lydall DA, Roberts JJ (1986) Cancer Res 46:1972
9. Canovese L, Cattalini l, Chessa G, Tobe ML (1988) J Chem Soc (Dalton Trans) 2135
10. Qu Y, Bloemink MJ, Mellish KJ, Rauter H, Smeds KA, Farrell N (1997) Factors Affecting Formation and Structure of DNA Intrastrand Cross-links by Dinuclear Platinum Complexes. In: Hadjiliadis N (ed) Cytotoxic, Mutagenic and Carcinogenic Potential of Heavy Metals Related to Human Environment NATO ASI 26:435 Series 2
11. Broomhead JA, Rendina LM, Sterns M (1992) Inorg Chem 31:1880
12. Schumann E, Altman J, Karaghisoff K, Beck, W (1995) Inorg Chem 34:2316
13. Comess KM, Lippard SJ (1993) Molecular Aspects of platinum-DNA interactions. In: Neidle S, Waring M (eds) Molecular Aspects of Anticancer Drug-DNA Interactions. CRC Press, Boca Raton FA, Vol. 1 p 134

14. Chu G (1994) J Biol Chem 269:787
15. Farrell N (1995) Comments in Inorganic Chemistry 16:373
16. Farrell N, Qu Y, Feng L, Van Houten B (1990) Biochemistry 29:9522
17. Van Houten B, Illenye S, Qu Y, Farrell N (1993) Biochemistry 32:11794
18. Buning H, Altman J, Beck W, Zorbas H (1997) Biochemistry 36: 11408
19. Qu Y, Farrell N, (1995) Inorg Chem 34:3573
20. Skov KA, Adomat H (unpublished observations)
21. Milkevitch M, Storrie H, Brauns E, Brewer KJ, Shirley BW (1997) Inorg Chem 36:4534
22. Milkevitch M, Shirley BW, Brewer KJ (1997) Inorg Chim Acta 264:249
23. Sabat M (1996) Metal Ions Biol Sys 32:521
24. Chu G (1994) J Biol Chem 269:787
25. Zlatanova J, Yaneva J, Leuba SH (1998) The FASEB Journal 12:791
26. Auge P, Kozelka J (1997) Transition Met Chem 22:91
27. Roberts JD, van Houten B, Qu Y, Farrell N (1989) Nuc Acids Res 17:9719
28. Hills CA, Kelland LR, Abel G, Siracky J, Wilson AP, Harrap KR (1989) Brit J Cancer 59:527
29. Startford IJ, Workman P (1998) Anti-cancer Drug Design 13:519
30. Farrell N (1989) Prog Clin Biol Med 10:90
31. Matthews J, Adomat, H, Farrell N, King P, Koch C, Lord E, Palcic B, Poulin N, Sangulin J, Skov KA (1996) Br J Cancer 73:S200
32. Matthews JB, Adomat H, Skov KA (1993) Anti-Cancer Drugs 4:463
33. Skov KA, Adomat H, Farrell N, Matthews JB (1998) Anti-Cancer Drug Design 13:207
34. Soulie P, Raymond E, Misset JL, Cvitkovic E (1996) Oxaliplatin: Update on a Active and safe dach platinum complex In: Pinedo HM, Schornagel JH (eds) Platinum and Other Metal coordination Compounds in cancer Chemotherapy 2. Plenum, New York, p 165
35. Micetich KC, Barnes D, Erickson LC (1985) Cancer Res 45:4043
36. Farrell N (1996) Metal Ions in Biol Sys 32:603
37. Farrell N, Qu Y, Hacker MP (1990) J Med Chem 33:2179
38. Farrell N, Roberts JD, Qu Y, Zou Y, Marples B, Skov KA, Tognella S (1994) Proc. AACR 35:2637

Role of Metal Ions in Antisense and Antigene Strategies

Bernhard Lippert[1], Marc Leng[2]

Fachbereich Chemie, Universität Dortmund, D-44221 Dortmund, Germany
Centre de Biophysique Moléculaire, CNRS, Rue Charles-Sadron, F-45071 Orléans Cedex 2, France
E-mail: [1] lippert@pop.uni-dortmund.de, [2] leng@cnrs-orleans.fr

The fields of gene regulation by antisense and antigene agents and the therapeutic use of oligonucleotides in general are developing at a rapid pace. The present status of this research and the various approaches are briefly reviewed. Specific attention is given to the role of metal ions in these processes and in particular to the potential usefulness of platinum am(m)ine complexes as site-directed oligonucleotide cross-linking agents.

Keywords. Antisense, Antigene, Oligonucleotides, Metal ions, Platinum

1
Introduction

The idea to regulate gene expression by the use of deoxyoligonucleotides was first convincingly demonstrated when in 1978 Zamecnik and Stephenson showed that a synthetic 13-mer complementary to the 3'- and 5'-terminal sequences of the Rous Sarcoma virus 35 S RNA was able to inhibit virus production in infected chick embryo fibroblast cells [1]. Later, Zamecnik et al. were able to inhibit viral replication and expression of human T-cell lymphotropic virus type III in cultured cells by adding a 20-mer deoxyoligoribonucleotide directed against a splice acceptor site of the viral RNA [2]. Ever since the interest in potential applications of oligonucleotides as chemotherapeutic agents has been high, with a number of related concepts – antisense strategy [3], antigene approach [4], ribozyme targeting [5], aptamer binding to proteins [6] – presently being investigated and in clinical testing. Despite considerable obstacles for in vivo application – e.g. problems of cell delivery, problems of oligonucleotide stability, proper choice to target sequence – there is great optimism as far as the long-term outlook for the successful use of these concepts is concerned [7]. This optimism seems to be justified considering the fact that it is now well established that nature itself, at least in prokaryotes, employs the antisense strategy to control gene activation and expression, probably by using short RNA molecules as repressors of gene expression [8]. In accordance with this feature, emerging clinical data on the usefulness of antisense therapeutics in the treatment of viral diseases (e.g. human papillomavirus, cytomegalovirus, HIV), certain cancers (e.g. acute or chronic myelogenous leukaemia), as well as of various kinds of inflammatory processes (e.g. Crohn's disease, psoriasis, ulcerative colitis), are very promising [9,10].

2
Molecular Basis for Gene Regulation by Oligonucleotides

DNA is replicated, transcribed into mRNA, and the mRNA is translated into proteins (Scheme 1). Usually DNA is the genetic material but if it is RNA, such as in RNA retroviruses, RNA is initially converted into DNA by reverse transcriptase before being processed in the usual way. Gene expression, viz. formation of a

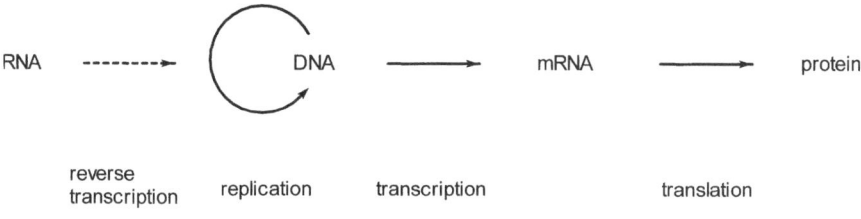

Scheme 1

protein, can be inhibited at any of these stages, and oligonucleotides can be applied to interfere with these processes.

2.1
Preventing Translation: Principles of Antisense Approach

Single-stranded DNA oligonucleotides that hybridize with a mRNA target strand through Watson-Crick base-pair formation, i. e. in an antiparallel fashion, are capable of blocking translation [9a]. In a similar fashion short RNA molecules lacking coding capacity, yet being complementary to mRNA, can hybridize with the latter and repress translation [8]. Due to their general instability towards enzymatic degradation, RNA antisense molecules, despite their importance in living systems, appear not to be ideal for therapeutic purposes (see, however, Sect. 2.3).

As will be pointed out in Chap. 3, a length of 10–20 bases in the oligonucleotide is usually sufficient to achieve translational arrest. There are at least two scenarios that can explain the effect: (1) the artificial oligonucleotide impairs the function of the ribosome by steric blockage, leading to an incomplete protein and a loss of protein function, and (2) the ubiquitous enzyme RNaseH is activated and degrades the mRNA strand complementary to the antisense DNA oligomer at the site of hybridization (Scheme 2). While the latter mechanism is generally believed to be the most important one responsible in vivo, the first possibility might be operative in cases where a chemically modified antisense oligonucleotide has become RNaseH-resistant or if a cell is deprived of this enzyme.

2.2
Impairing Transcription (and Replication): Antigene Approach

Control of gene expression at the DNA level is to be considered the ultimate goal, in view of the fact that each cell contains only a single copy of DNA as opposed to many copies of mRNA when transcription is functioning. Inhibition of DNA transcription by a deoxyoligonucleotide was indeed demonstrated in an in vitro

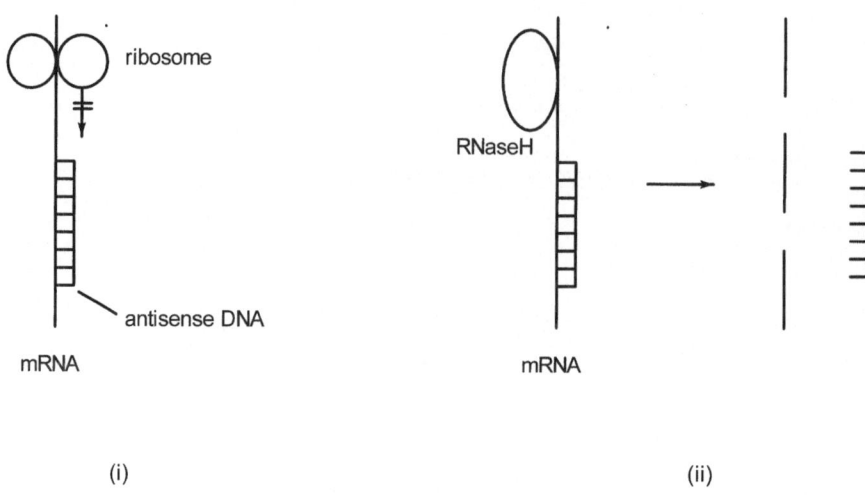

Scheme 2

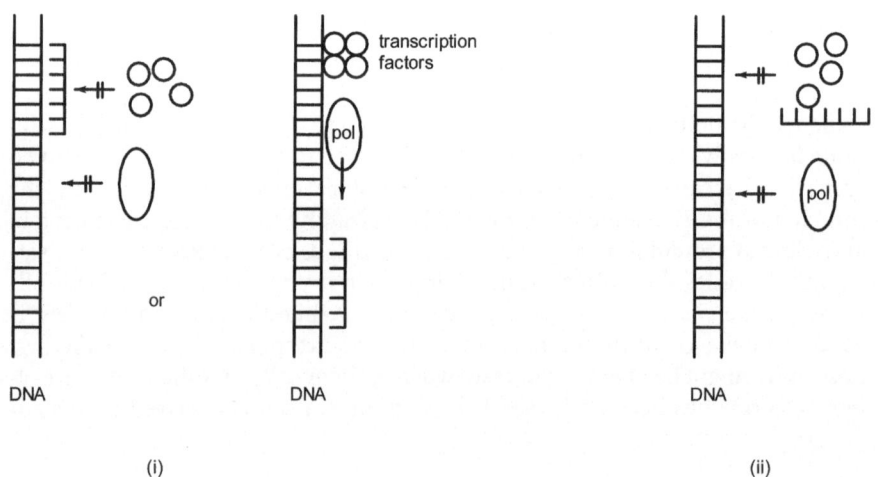

Scheme 3

experiment in 1988 by Cooney et al. [11]. Again, there are (at least) two explana-
tions for this effect (Scheme 3).
1. Formation of a local DNA triplex. Nucleic acid triplex structures have been
   known since the 1950s, and in 1987 Moser and Dervan [12], as well as Hélène

and co-workers [13], convincingly demonstrated that this scenario is likely to account for gene silencing. Depending on the site of triplex formation, either the initiation or the elongation process of mRNA formation can be affected. In the first case binding of transcription factors may be inhibited, whereas the second case could be the consequence of downstream triplex formation.
2. The oligonucleotide may act as a decoy for transcription factors, thereby interfering with mRNA formation. However, in general, mutagenesis studies seem not to support such an idea [14].

Unlike in the case of RNaseH-mediated antisense destruction of mRNA, where a long residence time at the target site appears not to be necessary, the antigene approach requires the oligonucleotide to bind strongly and long-lastingly to duplex DNA. Routes applied to reach this goal are the use of conjugates of the oligonucleotide with intercalators or of triple helix-specific intercalating ligands [15], minor groove binders [16], or simply by reducing the negative charge of the oligonucleotide backbone and by substituting the phosphodiester entities by neutral [17], or cationic [18] linkers (cf. also Sect. 3.3). Alternatively, oligonucleotides tethered with groups capable of forming covalent bonds with the target DNA, hence with the ability for cross-linking the third strand and DNA, have

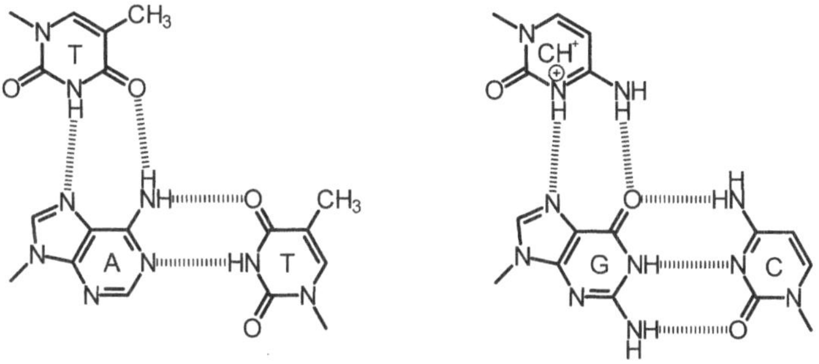

Scheme 4

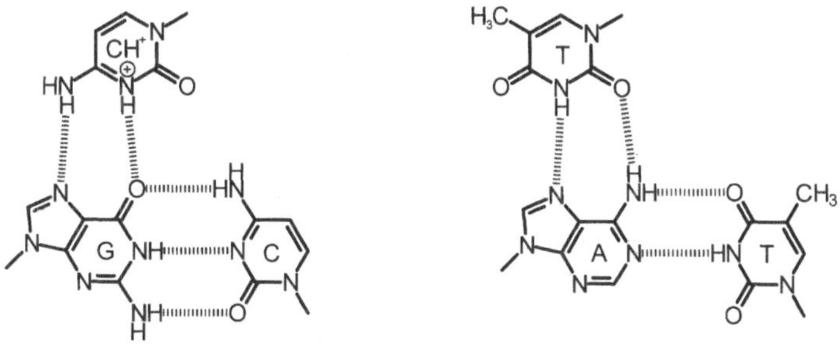

Scheme 5

also been studied [19]. Among these, attempts have been made to recruit the ubiquitous mammalian enzyme Topoisomerase I (TopoI) to cleave DNA by tethering a TopoI inhibitor, camptothecin, to the third strand and "attracting" TopoI to cleave one strand of the double-strand DNA, at the site determined by the oligonucleotide sequence [20] (see also Sect. 3.4).

Details of base triplet geometries, viz. H-bonding patterns, glycosidic bond orientation (*anti* or *syn*), strand direction (parallel or antiparallel) etc., are reasonably well understood [21]. They were obtained from a combination of different methods such as theoretical calculations [22], NMR spectroscopy [23], X-ray analysis [24], molecular modeling [25], etc. In the simplest case a Watson-Crick AT pair hydrogen bonds with another T in the Hoogsteen fashion, and a Watson-Crick GC pair does so with a protonated CH^+ (Scheme 4). As far as triple helices are concerned, this requires a purine strand of a double-stranded DNA to interact with an oligonucleotide consisting exclusively of pyrimidine bases. In that case the third (pyrimidine) strand runs parallel to the homopurine strand, with all nucleotides in the *anti* conformation. However, if the interaction of the third pyrimidine base with the Watson-Crick pairs is of the reversed-Hoogsteen type (Scheme 5), the third strand is antiparallel to the homo purine strand.

The strand direction of the third strand is reversed if the nucleotides of this strand adopt *syn* conformations. Purine bases in the third strand can likewise be applied [4] but rules for the purine,purine-pyrimidine family of triplexes are more complex. Many efforts are presently being devoted to overcoming limitations in generating universally applicable oligonucleotide sequences for DNA triplex formation [26].

2.3
Ribozymes as Cleaving Agents of mRNAs

The ribozyme concept tries to take advantage of naturally occurring self-splicing RNA molecules which either by hydrolysis or transesterfication cleave themselves or other RNA sequences. A number of ribozyme motifs are known today, which include "hammerhead", "hairpin", "hepatitis delta virus", as well as other types. The suitability of ribozymes as therapeutic agents stems from the fact that they can be directed toward mRNA transcripts of disease-related overexpressed proteins (Scheme 6) [27]. Two approaches are currently being pursued:

1. Delivery of chemically modified ribozymes by "conventional" methods (e.g. via liposomes). In this approach chemical modification is necessary in order to avoid rapid degradation by RNases.
2. Delivery by retroviral vectors. In this case the artificial gene for the ribozyme is introduced into the genome ("gene therapy") and expressed over a long time [28]. This approach is particularly useful in systemic diseases such as AIDS.

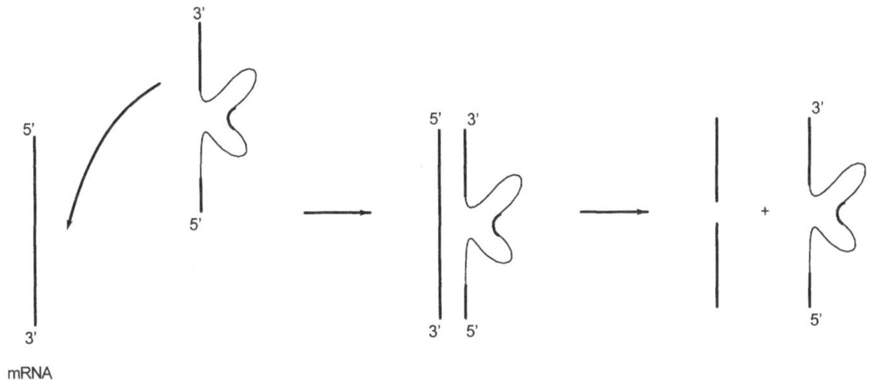

mRNA

Scheme 6

2.4
Aptamers for Transcription Factor or Enzyme Inactivation

Aptamers [29] are oligoribo- or oligodeoxyribonucleotides which bind with high affinity to proteins [29a,b] or recognize specific DNA structures, e.g. a hairpin element [29c]. The mechanism by which these nucleic acid molecules work is unrelated to the antisense or antigene mechanism. Rather a vital protein, e.g. a transcription factor responsible for initiation of DNA transcription, or a crucial enzyme, is inhibited. Aptamers are usually identified from combinatorial libraries and in vitro selection (SELEX technology). Very little is known about details of aptamer oligonucleotide-protein interactions although general principles of nucleic acid-protein interactions are expected to be followed. Moreover, the fact that several known aptamers, active against completely different proteins, contain guanine stretches that fold into unimolecular quadruplex structures [30], adds to the puzzle of their mode of action.

2.5
Double-Stranded RNA as a Gene Inhibitor

The most recent addition to the intriguing class of gene-inhibiting nucleic acids is double-stranded RNA. For reasons as yet unknown microinjection of dsRNA complementary to the coding region of a certain gene, yet not of (single-stranded) antisense RNA, into the nematode worm *C. elegans* causes a specific protein production to be shut down. The extraordinary low concentrations – few copies of the dsRNA per cell – represent a major puzzle [31].

3
Requirements for Oligonucleotides

3.1
Oligonucleotide Length for Antisense Approach

It is generally assumed that an oligonucleotide containing 12–15 nucleobases is sufficient to target a unique sequence of an mRNA. This figure is obtained as follows [32]: Of the mammalian (haploid) genome (ca. 3×10^9 base pairs) only an estimated 1–3% is expressed in any given cell, meaning that the genome that actually codes for proteins amounts to $3 \times 10^7 – 1 \times 10^8$ base pairs. If it is taken into account that due to the complicated tertiary structure of mRNA and protein association only 10% are single-stranded regions sufficiently long (8–12 bases) to allow hybridization, in essence $3 \times 10^5 – 1 \times 10^7$ bases represent potential targets. With the number of unique sequences being a combination of the four bases, it follows that a 12 mer of specific sequence, according to probability considerations is present only once in a number of $4^{12} = 1.68 \times 10^7$ bases.

Today, it is accepted that even shorter oligonucleotides (n<10) may be useful, especially if they have a high degree of conformational fit ("pre-ordered" oligonucleotides) toward the target mRNA as a result of appropriate chemical modification. This feature ensures a high rate constant for hybridization, a feature possibly as equally important as the binding strength. Efforts have also been made to account for the highly irregular structures (single strands, double-stranded regions, loops, bulges, pseudoknots etc.) of RNAs by constructing two short oligonucleotides separated by a flexible synthetic tether, which are capable of binding to spatially removed regions of the RNA target [33].

3.2
Oligonucleotide Length for Antigene Approach

In the antigene approach, that is triplex formation, usually longer oligonucleotides (n>15) are applied in order to ensure tight binding. Considering the complicated structure of chromatin, triplex formation may appear rather unlikely, but it has to be taken into consideration that any DNA region can only be transcribed if it is accessible to the surrounding medium.

3.3
Oligonucleotide Stability; Modifications

Oligonucleotides applied as therapeutic agents need to be stable to nucleases which degrade them. Therefore much synthetic work is being devoted towards this goal and numerous modifications have been made at the oligonucleotide backbone, the sugar and the heterocyclic part of the bases [34]. As has been found, virtually any change made to the phosphate backbone confers stability. As far as clinical applications are concerned, the phosphorothioates [35], that is

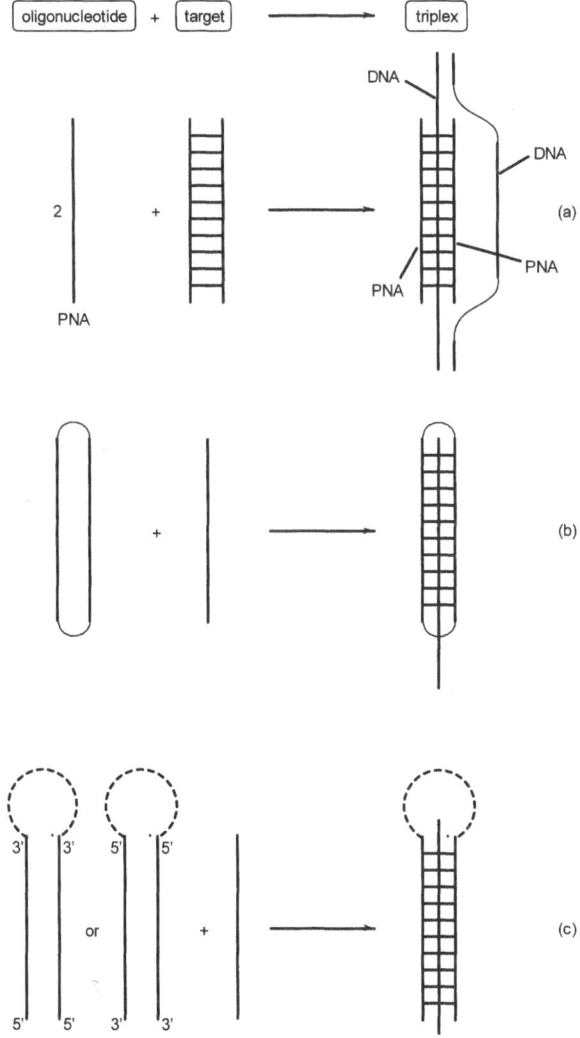

Scheme 7

oligonucleotides in which a phosphate-bound oxygen atom is replaced by a sulfur atom, have been the most successful so far. However, the phosphorothioates are also known to interact in a non-specific way with other biomolecules. Very encouraging also are tests which have been carried out with oligonucleotides modified at the 2' position of the sugar (2'-O-methyl or 2'-methoxyethyl phosphate) [10].

There are numerous examples of oligonucleotide analogues in which the sugar-phosphate backbone has been entirely replaced by other entities, whether negatively charged as the natural oligonucleotides, neutral, or positively charged. Among these, the Peptide Nucleic Acids (PNAs) have been the most studied [36]. Binding of PNAs containing exclusively pyrimidine bases to double-stranded DNA results in strand replacement, that is two PNA molecules and one DNA strand form an extremely stable triplex, while the second DNA strand becomes unpaired and is then susceptible to single-strand-specific nucleobases as, for example, in Scheme 7a. In a similar way, circular oligonucleotides [37] or so-called clamp oligonucleotides [38] have been synthesized, which form triplexes with single-stranded nucleic acid target strands (Scheme 7b,c). In the "clamp oligonucleotides", an artificial linker joins identical ends, e.g. 3' or 5', depending on the desired strand direction in the product.

3.4
Tethering Reactive Groups

Conjugates between oligonucleotides and reactive entities have been synthesized in large numbers. In all cases the idea is to either irreversibly cross-link the oligonucleotide and the target or, via a secondary reaction, to destroy the target. Photocross-linking [39] and cross-linking by an alkylating electrophile [40] are among the most widely applied approaches. Metal ions attached to an oligonucleotide and capable of cleaving target strands either hydrolytically or through redox chemistry have likewise been used and will be discussed separately (cf. Sects. 4.2 and 4.3). In vitro efficiency has been demonstrated in many cases, but in vivo application is more problematic as there is the potential of undesirable toxicity as a consequence of the reactive group.

4
Involvement of Metal Ions

4.1
Stabilization of Multistranded Nucleic Acids by Metal Ions

Relatively little is known if and how metal ions influence the gene-regulating processes outlined above, even though it is generally agreed that di- or multiple-stranded oligonucleotides require cations for charge neutralization [41] (unless backbone modification has rendered them neutral or positively charged). Metal-ion binding to either phosphate oxygen atoms or nucleobase donor sites has been reported for almost all three – and tetrastranded – nucleic acid structures, be it inter- [42] or intramolecular DNA triplexes [43], or four-stranded structures (quadruplex DNA [44] or RNA [45]; aptamers [29a]; telomeres [46]; Holliday junctions [47]). The metal ions involved are usually the intracellular bulk metals Mg^{2+} and K^+, but in many cases the exact positioning of the metal ions is as yet unclear. An exception represents the situation in guanine quartets present among others in telomeres or certain aptamers, for which K^+ binding to eight O6 sites of two adjacent guanine quartets has been established [29a,44–46].

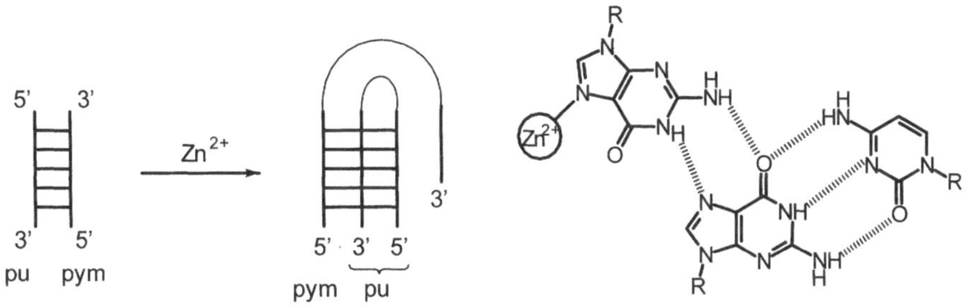

Scheme 8

The effect of other metal ions, e.g. of a trace element such as Zn^{2+} or of other exogenous metal ions including such of known toxicity, is of substantial interest [48]. For example, it has been demonstrated that the B-DNA fragment $d(GA\cdot CT)_{22}$ in the presence of Zn^{2+} spontaneously adopts an intramolecular triplex structure consisting of a homopurine, homopurine, homopyrimidine type with an unpaired homopyrimidine strand left [49] (Scheme 8). Hydrogen bonding between the purine strand of the duplex and the third purine strand involves G,G and A,A pairs. The role of Zn^{2+} has been proposed to be that of binding to the N7 site of the purine base in the third strand, thereby neutralizing the phosphate group and at the same time polarizing the purine base in such a way that H-bonding to the duplex is reinforced [50].

4.2
Metal Ions and Ribozymes

It is well established that phosphate ester hydrolysis and phosphoryl transfer are strongly metal-ion dependent and responsible for the observed rate acceleration at neutral pH when compared to metal-free conditions. The role of the metal ion or ions is to provide the nucleophile (coordinated hydroxide) for attack of the phosphorus atom, to act as a Lewis acid by polarizing P-O bonds and making the P more susceptible to attack, and/or possibly to stabilize the leaving group [51]. In addition, metal ions also have the function of accomplishing the proper folding of the RNA molecule, similar to the situation of tightly bound Mg^{2+} ions in tRNAs. The specific role of divalent metal ions in ribozyme functions has represented a major focus of research for several years. It is still not fully understood and is sometimes controversially discussed [52]. Among others phosphorothioate backbone modification, metal ion specificity, and inhibition by "wrong" metal ions have been used to improve the understanding of the various functions of metal ions. In the case of the "hammerhead" ribozyme, two independent X-ray crystallographic structure analyses have revealed tight binding of divalent metal ions to six sites [53]. Time-resolved X-ray crystallography [53a] seems to support the idea that a single Mg^{2+} ion bound to a phosphate-oxygen atom (*pro-R*) in the catalytic pocket acts as a Lewis acid, and at the same time provides the hydroxo ligand responsible for deprotonation of the 2'-OH group of the right C residue and subsequent attack of 2'-O$^-$ on P.

4.3
Site-Directed Artificial Nucleases

The principles of antisense or antigene recognition between an oligonucleotide and a target sequence can be exploited by directing cleaving agents to a specific site in a large RNA or DNA molecule and to cut these nucleic acids in vitro. There are two ways to achieve this, either by redox chemistry or via hydrolysis.

In the first case, a redox-active metal ion is attached via a linker to the 5' end, the 3' end or both ends of the oligonucleotide. The most widely applied metal entities are EDTA-Fe^{II}, as pioneered by Dervan [12,54] and o-phen-Cu^{I}, introduced by Sigman [55], with several other approaches, e.g. cationic manganese porphyrin groups [56] also in use. Once delivered to the target strand(s), a coreactant, such as H$_2$O$_2$ or another peroxo compound, is added and leads to OH radicals or a reactive metal species in a higher oxidation state. It was in fact an experiment of this type – delivery of EDTA-Fe tethered to the 5' end of a pyrimidine base 15-mer and directed toward a stretch of complementary purine bases within a DNA molecule containing more than 4000 base pairs – which in 1987 provided proof for the antigene approach [12]. Meanwhile this principle has been verified many times and has been successfully applied to much larger DNA molecules, among others also to the short arm of the human chromosome 4 which is comprised of more than 200 mega base pairs [54b]. Exact complementarity between the oligonucleotide and the target sequence is absolutely crucial. Even a single base mismatch leads to a dramatic loss in cleavage efficiency. It is of interest to note that in favorable cases it is possible to gain insight into the location of the oxidizing species (major or minor groove) on the basis of the nature of the different sugar cleavage products.

The second principal way utilized in nucleic acid cleavage reactions is through backbone hydrolysis. While relatively easy in the case of an RNA target, this approach is exceedingly difficult in the case of DNA – unless metal ions come into play. It has been estimated that the half-life of a phosphodiester bond at pH 7 at 25 °C is 200 million years. Metal ions, in particular if acting in a concerted fashion, can accelerate this process enormously [57]. The concept of site-directed hydrolysis of nucleic acids by metal-assisted catalysis is based on the idea of attaching a suitable metal entity via a synthetic tether to an oligonuleotide and to use the recognition process of the oligonucleotide to find the target [58]. In related strategies, conjugates between double-stranded DNA molecules and a metal entity have been constructed to cleave RNA via triplex formation [59], and natural metal-dependent (Ca^{2+}) nucleases have been attached to oligonucleotides to cleave the target [60]. Among the metal ions presently of particular interest, as far as in vitro studies are concerned, are ions of the lanthanides and combinations thereof, e.g. Ce^{IV} and Pr^{III} [61], which appear to react synergistically and cleave DNA at an unprecedented rate. The introduction of these ions in suitable linker entities represents a major challenge to preparative chemistry.

5
Platinated Oligonucleotides

Complex formation between cationic Pt^{II} am(m)ine species and polyanionic nucleic acids has been a major research area over the last 30 years following the discovery of cis-$Pt(NH_3)_2Cl_2$ (cisplatin) as a powerful antitumor agent, with DNA being the likely ultimate target [62]. As a result of this work it has been recognized that DNA cross-linking adducts of cisplatin, and also those of the corresponding trans-isomer (transplatin) are generally stable and in most cases also kinetically inert. There are, however, a few notable exceptions [63]. The strengths of Pt-N (nucleobase) bonds are usually sufficiently high to cause quite substantial steric distortion of DNA or oligonucleotides, even though the Pt coordination geometry seems to suffer likewise [64]. The present view is that DNA distortion as a consequence of intrastrand DNA cross-linking triggers a complex reaction cascade which involves a number of players (protein factors) and eventually results in cell death of the tumor cell.

5.1
Pt Cross-Linking of Oligonucleotides: Early Work

Studies aimed at a better understanding of the effects of cis- or transplatin on DNA and oligonucleotides have been conducted in large numbers but will not be reviewed here. Rather, two approaches with the aim of systematically cross-linking oligonucleotide strands by bifunctional Pt reagents will be referred to. Vlassov et al. [65] applied $[BrPt(dien)-(CH_2)_6-(dien)Pt(H_2O)]^+$ (dien=diethylenediamine) to attach this reagent first to a guanine nucleobase in the center of a DNA single strand and subsequently to make a cross-link with a complementary strand, also via guanine, yet several bases removed from the former (Scheme 9a). Base pairing via Watson-Crick recognition appeared to be crucial for the success of this reaction in that non-complementary strand cross-linking proved rather inefficient.

Orgel et al. [66] studied cross-linking reactions between phosphorothioate oligonucleotides and single-stranded DNA and RNA molecules by cis- and trans-$Pt(NH_3)_2Cl_2$ as well as K_2PtCl_4. Their findings confirmed earlier results on the affinity of the S-groups in the backbone for Pt^{II}[67] and at the same time showed cross-linking to A (adenine), G (guanine) or T (thymine) base residues in a bulge region near the opposite end of the target sequence (Scheme 9b).

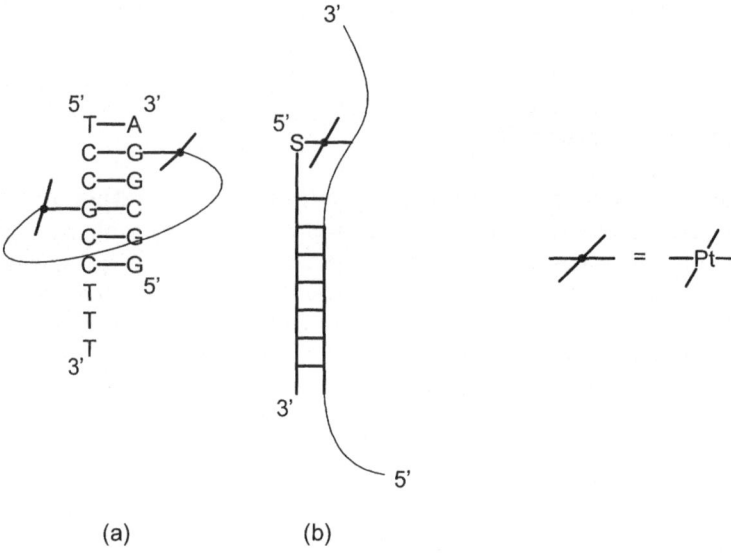

(a)   (b)

Scheme 9

5.2
Interstrand DNA Cross-Linking by Transplatin: Models

Unlike cisplatin, which preferentially forms intrastrand adducts between neighboring guanine bases or between adenine and guanine (N7 positions each) and only occasionally interstrand cross-links [68], transplatin forms interstrand cross-links more frequently, and in particular between complementary nucleobases [69]. As part of a systematic study of structural aspects of such adducts we prepared a series of model compounds of transplatin with model nucleobases [63b], including metalated analogues of G and C (cytosine) as well as A and T bases with Watson-Crick [70], reversed Watson-Crick [71], Hoogsteen [70,72], and reversed Hoogsteen [73] arrangement of the bases. Some of the compounds prepared and characterized are of interest in themselves for other reasons such as spontaneous self-aggregation to give platinated nucleobase quartets [73,74]. As for Watson-Crick and Hoogsteen arrangements of the cross-linked bases, they can be considered models of putative products of antisense reactions between a platinated oligonucleotide and a suitable target (Scheme 10).

The geometry of the G,C cross-link should correspond to that of a transplatin-modified DNA, as obtained from molecular modeling [75] and NMR spectroscopy [76]. While results of this work suggest a rise in base stacking as a consequence of the steric bulk of the NH_3 groups at the Pt, this feature may not be mandatory; however, consider findings in crystal structure determinations of other bis(nucleobase) adducts of *trans*-$(NH_3)_2Pt^{II}$, where intermolecular H-

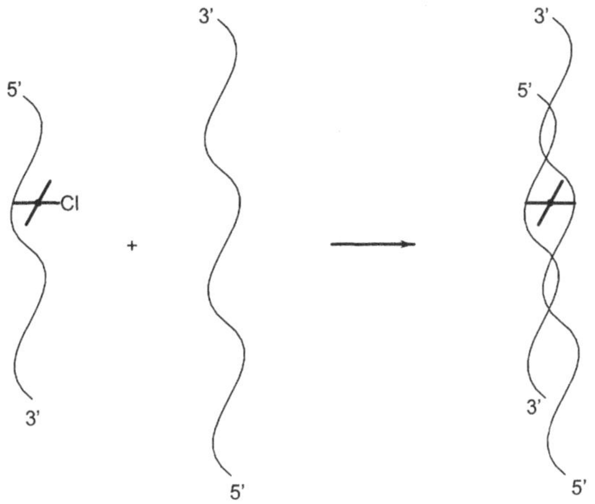

Scheme 10

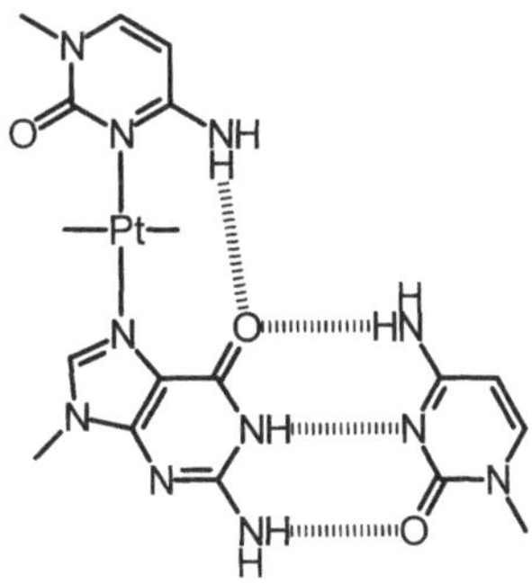

Scheme 11

bonding between the NH_3 groups and nucleobase oxygen acceptors leads to a remarkable tilting of the Pt-NH_3 vectors toward the bases, leaving the stacking interaction virtually unaffected [77].

5.3
Model for a Platinated Triplex

The idea to apply trans-$(NH_3)_2Pt^{II}$-modified oligonucleotides as potential antigene agents came from a model compound, obtained upon cocrystallizing a Hoogsteen-modified GC pair with free C (Scheme 11) [78]. The geometry of the complex is remarkably similar to that of the $CH^+=G\equiv C$ triplet present in triple-stranded DNA (cf. Sect. 2.2). Unlike the approaches used by Vlassov and Orgel (see Sect. 5.1), Pt cross-linking in this case was exclusively via donor sites of the heterocyclic part of the nucleobases involved, leading to a minimum of flexibility of the Pt entity and in a way to a "pre-order" of the reacting entity.

5.4
From Models to Oligonucleotides: Platination Reactions

We are pursuing different strategies to prepare site-specific $trans$-$(NH_3)_2Pt^{II}$-modified DNA oligonucleotides.

Direct Platination. Reactions of $trans$-$[Pt(NH_3)_2Cl_2]$ or $trans$-$[Pt(NH_3)_2Cl(H_2O)]^+$ with oligonucleotides, containing exclusively the pyrimidine bases T and C, occur selectively at cytosine-N3, but reactions are slow and give, with several cytosine bases present, a mixture of differently platinated species [79]. The latter situation is not necessarily a disadvantage in that all species, in principle, are reactive. Platination reactions are substantially accelerated if a single guanine base is introduced in the pyrimidine oligonucleotide [80] and products can be purified by HPLC and characterized by ^{1}H NMR spectroscopy and electrospray ionization mass spectrometry. Suicide reactions, viz. formation of long-range internal cross-links, can occur but generally are not a major problem.

Use of a Protective Group. In order to exclude undesired side reactions (inter- or intrastrand cross-linking) of the reactive chloro group in complexes of type (oligonucleotide)Pt$(NH_3)_2$Cl prior to binding to the target, we used acid-labile protective groups L at the Pt, viz. used $trans$-Pt$(NH_3)_2$LCl species in platination reactions of oligonucleotides. As an L group we applied the 1-methylthyminate anion, which binds to Pt via the N3 position and protects the Pt efficiently from nucleophilic attack. Only upon protonation is cleavage and activation of the Pt center achieved (Scheme 12). We have demonstrated [79] that the reaction works, but that it is only applicable in the absence of purine-containing oligonucleotides, otherwise depurination, especially of A residues, takes place. Clearly, protective groups that can be removed under milder conditions are desirable.

Automated Synthesis. We have previously demonstrated that the preparation of site-specifically platinated DNA via automated synthesis and application of nu-

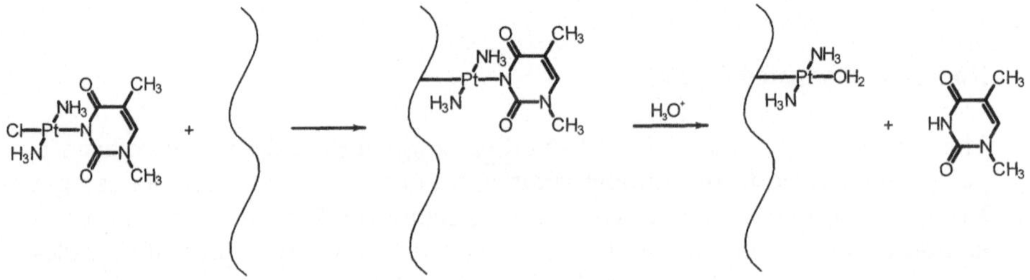

Scheme 12

cleoside H phosphonate methodology is in principle possible [81]. The problem not yet solved is the protection of the monofunctional Pt adduct within the oligonucleotide from reacting with N donor compounds (pyridine or NH_3) present during oligonucleotide synthesis, which render the Pt moiety unreactive for cross-linking with a target molecule. Similarly, solubility problems of the platinated base building blocks can present a problem.

5.5
Activating the Platinated Strand in the Presence of the Target

As pointed out above, keeping a monofunctionally bound $trans$-$(NH_3)_2Pt^{II}$ entity in a reactive state for cross-linking is a major challenge. An ideal situation is realized if activation of a protected Pt species is exclusively triggered by the presence of a target, e.g. during hybridization and formation of a double helix. In two instances such reactions have been observed. One case deals with the bifunctional 1,3-$trans$-$\{Pt(NH_3)_2[GNG]\}$ intrastrand cross-links where N is a nucleotide residue and the other with the cisplatin monofunctional adducts cis-$[Pt(NH_3)_2 (G)(het)]^{(n+1)+}$ where het is an heterocyclic amine such as pyridine, 5-nitroquinoline, N-methyl-2,7-diazapyrenium, ellipticine, etc.

Rearrangement of 1,3-$trans$-$\{Pt(NH_3)_2[GNG]\}$ Cross-Links. In the reaction between native DNA and transplatin, at low level of platination, monofunctional adducts are rapidly formed and then they slowly react further to form mainly interstrand cross-links between complementary G and C residues [69]. In the reaction between transplatin and single-stranded oligonucleotides containing the triplet GNG 1,3-$trans$-$\{Pt(NH_3)_2[GNG]\}$, intrastrand cross-links are formed [82,83]. As long as the platinated oligonucleotides are single-stranded, the bifunctional adducts are stable even in the presence of NaCl. There is one exception concerning the stability of the adducts in which the base residue adjacent to the adducts at the 5' side is a C (CGNG). In this case the metal migrates from the 5' G to the 5' C residue and an equilibrium between the two isomers (G1,G3- and C1,G4-intrastrand cross-links) is attained [84,85]. The reaction is rather slow ($t_{1/2} \sim 38$ h at 37 °C). The rearrangement does not occur when the C residue is on the 3' side of the adduct.

Hybridization of the oligonucleotides containing a 1,3-intrastrand cross-link with the complementary sequences promotes the rearrangement of the intrastrand cross-link into an interstrand cross-link [86]. A schematic representation of the linkage isomerisation reaction is given in Scheme 13. N and N' are com-

Scheme 13

plementary nucleotide residues. The mechanism of the reaction has not yet been completely elucidated but some proposals have been made [85–88].

The reaction does not seem to proceed through a solvent-associated intermediate. The interstrand cross-link is always formed between the platinated 5' G and the complementary C, and never between the platinated 3' G and the complementary C. The cleavage of the Pt-G(3') bond with formation of a monofunctional adduct is excluded by trapping experiments and thus a direct nucleophilic attack of the Pt-G(3') bond by the C complementary to the G(5') is suggested. This implies that the attacking C residue is located near the platinum residue and along its Z-axis. The platinated duplex is distorted at the level of the adduct [89]. It is partially denatured, unwound (26°) and the helix axis is bent (45°). Molecular modeling supports that the attacking C residue is in the right position [90].

No evidence has been found of a direct participation of the sugar residues and the phosphate groups in the linkage isomerization reaction [88]. The replacement of the intervening nucleoside residue between the two cross-linked G residues by a propylene link or of the phosphate groups by uncharged methylphosphonate groups has no effect on the rate of the reaction. However, the rate is greatly decreased when the two cross-linked G residues are no longer linked. This suggests that the constraints present in the platinated macrocycle weaken the strength of the Pt-G_3 bond more than that of the Pt-G_5 bond.

The double helices flanking the adduct interfere by maintaining a well-defined structure at the level of the adduct and imposing local constraints. Within the hybrid the replacement of the deoxyribo-strand complementary to the platinated strand by a ribo-strand results in a decrease in the rate of the rearrangement by at least a factor of ten. On the contrary, the replacement, within this strand, of the triplet opposite to the adduct by the doublet 5'-UA or 5'-CA results in a tenfold increase in the rate of the reaction. The rate can be even faster when the platinated deoxy-strand is replaced by a (2'-O-methyl-ribo)-strand (the rearrangement is accomplished within a few minutes). In all these hybrids the interstrand cross-link is formed between the G and A residues. Again, the location of the attacking residue (A and no longer C residue) is crucial. The rearrangement does not occur when the doublet 5'-UA is replaced by the doublet 5'-AU [87] (Scheme 14).

Rearrangement of *cis*-[Pt(NH$_3$)$_2$(G)(het)]$^{(n+1)+}$ Adducts. In the reaction between *cis*-[Pt(NH$_3$)$_2$(het)Cl]$^{n+}$ and single-stranded oligonucleotides at acidic pH, the monofunctional *cis*-[Pt(NH$_3$)$_2$(G)(het)]$^{(n+1)+}$ adducts are formed [91,92]. These

Scheme 14

adducts are kinetically inert as long as the platinated oligonucleotides are single-stranded. The pairing of the platinated oligonucleotides with their complementary strands promotes two reactions [91,93,94]. One reaction is the cleavage of the bond between the platinum and the G residue with the release of *cis*-$[Pt(NH_3)_2(het)(H_2O)]^{(n+1)+}$. This complex can further react with any of the G residues within the hybrid resulting in a monofunctional *cis*-$[Pt(NH_3)_2(G)(het)]^{(n+1)+}$ adduct. However, the cycle release and reformation of the adduct stops because of the other reaction. The other reaction also promoted by the formation of the hybrid is the cleavage of the bond between the platinum and het which generates a monofunctional adduct *cis*-$[Pt(NH_3)_2(G)(H_2O)]^{2+}$. This adduct can further react with the neighboring residues to form an intra- or interstrand cross-link. Interstrand cross-links are preferentially formed at GC.GC sites [95]. The yields of the two concomitant reactions (cleavage of the Pt-G and Pt-het bonds) depend on the conformation of the hybrids and the nature of het. The second reaction is favored when het is ellipticine or *N*-methyl-2,7-diazapyrenium.

5.6
Interstrand Cross-Linking Reactions in Triplexes

As already pointed out (cf. Sect. 5.3), model nucleobase chemistry has led us to also pursue the potential use of transplatin-modified oligonucleotides in the antigene approach. Although a systematic study has not yet been carried out, the first results show the potential interest of this approach in the case of triplexes.

Two oligonucleotides containing either a single *trans*-$[Pt(NH_3)_2(G)(Cl)]^+$ or *trans*-$[Pt(NH_3)_2(C)(Cl)]^+$ adduct have been studied [96,97]. They bind to the complementary duplexes and stand in the major groove of the duplexes in a parallel orientation with respect to the homopurine strands of the duplexes. Whatever the nature of the monofunctional adducts, the interstrand cross-links result from the reaction between the monofunctional adducts and mainly the complementary G residues in the homopurine strands. In a first approximation, the rate of the cross-linking reaction is independent of the location of the monofunctional adduct along the oligonucleotide chain. This has been clearly shown for the *trans*-$[Pt(NH_3)_2(G)(Cl)]^+$ adduct located at the 5' end, in the middle or close to the 3' end of the oligonucleotide. In the case of *trans*-$[Pt(NH_3)_2(C)(Cl)]^+$ the experiments are not so straightforward because of the difficulty in platinating a given C residue within an oligonucleotide containing several C residues. The rate of the cross-linking reaction depends on the nature of the monofunctional adduct, being faster with the *trans*-$[Pt(NH_3)_2(C)(Cl)]^+$ adduct than with the *trans*-$[Pt(NH_3)_2(G)(Cl)]^+$ (at 25 °C, the half-times ($t_{1/2}$) for disappearance of the monofunctional adducts are about 9 and 15 h, respectively).

The *trans*-$[Pt(NH_3)_2(C)(Cl)]^+$ adduct is in the right position to attack the complementary G residue whereas the *trans*-$[Pt(NH_3)_2(G)(Cl)]^+$ adduct in the *anti* conformation has first to adopt the *syn* conformation to be in the right position. The fact that the rate of the cross-linking reaction is independent of the location of the *trans*-[Pt(NH$_3$)$_2$(G)(Cl)]$^+$ adduct along the oligonucleotide is due to a local denaturation of the triplex at the level of the adduct. Although the platination of the G residue does not interfere directly with the hydrogen bonding of the oligonucleotide to the complementary strand it decreases the thermal stability of the triplex. Two opposite effects arise from the presence of the monofunctional adduct. A stabilizing effect is due to the positive charge of the adduct and formation of a new Pt-N(nucleobase) bond, a destabilizing effect is due to a steric hindrance between the ammine ligands of the platinum residue and the adjacent base residues.

The closure of the *trans*-[Pt(NH$_3$)$_2$(G)(Cl)]$^+$ adduct into an interstrand cross-link seems to proceed through a solvent-associate intermediate, at least in the case of the adduct located at the 5' end of the oligonucleotide. The rate of the cross-linking reaction is about five times slower in 0.5 M NaCl than in 0.5 M NaClO$_4$. Under the same experimental conditions, the rate of closure of the

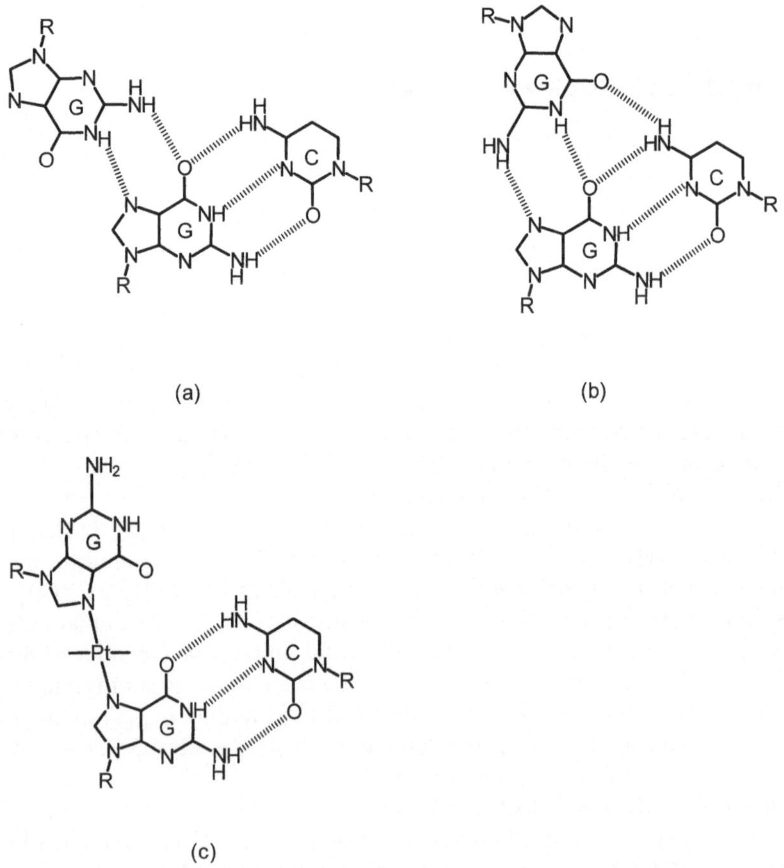

(a)   (b)

(c)

Scheme 15

trans-[Pt(NH$_3$)$_2$(C)(Cl)]$^+$ adduct is slightly reduced and thus one cannot exclude a direct displacement of chloride by the complementary G residue.

As far as the geometry of such adducts is concerned – C·G≡C and G·G≡C (with · representing a *trans*-a$_2$Pt^{II} entity) – both are feasible on the basis of model compounds: The first case, a Hoogsteen arrangement between C in the third strand (with the acidic proton replaced by the Pt moiety) and G of a Watson-Crick pair, has already been dealt with (cf. Sect. 5.3). The second case differs from the known GGC triples (Scheme 15a,b) [98] in that the interaction between the two guanines takes place, via Pt^{II}, exclusively through the N7 sites (Scheme 15c). X-ray crystallographic work on related model compounds [99] has clearly established that the situation sketched in Scheme 15c is realistic and permits an essentially coplanar arrangement of the two cross-linked guanines. The orientation of the glycosidic bond of the G in the third strand is such that a parallel orientation of the third strand (with respect to the homopurine strand of the duplex) is possible. The sugar of the third guanine would then be in the normal *anti* conformation.

6
Summary and Perspectives

The role that metal ions possibly play in the context of antisense and antigene approaches or, more generally, in cases where oligonucleotides are applied as therapeutic agents, has not been previously studied in great detail and for this reason is not very well understood. In the early days of antisense technology, an involvement of metal ions appears not to have been considered at all. On the other hand, the role of metal ions such as Mg^{2+} in stabilizing the tertiary structure of any RNA molecule is established, and the involvement of metal ions in ribozyme catalysis or in stabilizing guanine quartet structure in aptamers is now firmly established. Moreover, the use of metal-oligonucleotide conjugates as artificial nucleases in in vitro studies has proven to be extremely important in advancing this field.

There are other areas within this field where metal ions could be of significance, even though they are admittedly speculative at this moment. For example, the replacement of K$^+$ in aptamers by other suitable cations could further stabilize the guanine quartet core, and metal entities bound to an aptamer molecule might irreversibly cross-link the aptamer with a protein target molecule or simply reinforce the tertiary structure of the aptamer.

As outlined in Chap. 5, ourselves and others are focusing on the potential use of *trans*-Pta$_2$Cl$_2$ (a=NH$_3$ or other amine) as blocking agents in either antisense or antigene approaches. In order to obtain an antigene effect, undoubtedly a long-lasting blockage of DNA replication and transcription has to be achieved and if a non-RNaseH mechanism during antisense action is considered, the adduct has to be sufficiently long-lived, viz. it has to be sufficiently stable to avoid dissociation by the cellular machinery [3–5, 100]. As mentioned in Sect. 3.4, light activation and photo cross-linking is possible, in principle, but difficult to realize in in vivo experiments.

In this context, platinated oligonucleotides and specifically those containing transplatin G1,G3 cross-links, present some advantages. Unfortunately the rearrangement of the cis-$[Pt(NH_3)_2(G)(het)]^{(n+1)+}$ adducts into interstrand cross-links is too slow for in vivo applications. As to the 1,3-adducts of transplatin, they are stable as long as the oligonucleotides are single-stranded. The interstrand cross-linking reaction is only triggered by the formation of a double helix between the platinated oligonucleotide and the target. The reaction is fast and thus compatible with the life-times of most mRNAs. It is specific even in cells. The cross-linked oligonucleotides prevent the synthesis of proteins. Although the thermal stability of the hybrids is decreased by the G1,G3-intrastrand cross-links, this is not really a problem since chemical modifications of the oligonucleotide backbone increase the thermal stability of the hybrids without preventing the rearrangement of the intrastrand cross-links. In fact some of the chemical modifications favor it and, thus, relatively short oligonucleotides can be used. Moreover, these chemical modifications make the oligonucleotides resistant to nucleases [4,100] (cf. also Sect. 3.3). A major difficulty is to prepare these platinated oligonucleotides on a large scale for in vivo experiments. The yield of the reaction for the synthesis of the G1,G3-intrastrand cross-links is low (about 30%) in the favorable case of oligonucleotides containing a single GNG triplet. The presence of other G residues within the oligonucleotides decreases the yield dramatically. It is clear that the automated solid-phase synthesis of site-specifically platinated oligonucleotides [81,101] will facilitate the use of the platinated oligonucleotides in the antisense strategy and as a tool in molecular biology.

The pairing of an oligonucleotide containing either a transplatin G1,G3-intrastrand adduct or a cis-$[Pt(NH_3)_2(G)(het)]^{(n+1)+}$ adduct to a double-stranded DNA does not seem to induce the rearrangement of the adduct into an interstrand cross-link. The interstrand cross-linking reaction has, however, been achieved by using oligonucleotides containing either a single $trans$-$[Pt(NH_3)_2(G)(Cl)]^+$ or $trans$-$[Pt(NH_3)_2(C)(Cl)]^+$ adduct [96, 97]. Several improvements have to be achieved in order to use these platinated oligonucleotides in the context of the antigene strategy. A first point is to prevent the suicide reaction consisting of the formation of intrastrand cross-links within the free platinated oligonucleotides. Another point relates to the question of destabilization of the triplexes due to the monofunctional adducts. While metal cross-linking is expected to cause some steric distortion at the site of metal binding, this negative effect may, on the other hand, be outweighed by the strength of the metal-mediated bond between the strands. Finally, the rate of the cross-linking reaction is not very rapid, with $t_{1/2}$ in the range of a few hours. Work is in progress to circumvent these difficulties with the hope of eventually using platinated oligonucleotides for modulation of gene expression.

Acknowledgements. The Pt research mentioned was supported by the Deutsche Forschungsgemeinschaft and the Fonds der Chemischen Industrie (BL), by Centre National de la Recherche Scientifique and Agence Nationale de Recherches sur le SIDA (ML), and by BIOMED2 contract BMH4-CT 97–2485 (BL and ML). BL wishes to thank Jens Müller for preparation of the schemes and Susan Thompson, Matthias Janik and Markus Drumm for helpful comments and proof reading. ML thanks Dr. J.-M. Malinge for his comments.

References

1. (a) Zamecnik PC, Stephenson ML (1978) Proc Natl Acad Sci USA 75:280; (b) Stephenson ML, Zamecnik PC (1978) Proc Natl Acad Sci USA 75:285
2. Zamecnik PC, Goodchild J, Taguchi Y, Sarin PS (1986) Proc Natl Acad Sci USA 83:4143
3. (a) Stein CA, Cheng YC (1993) Science 261:1004; (b) Uhlmann E, Peyman A (1990) Chem Rev 90:543
4. Thuong NT, Hélène C (1993) Angew Chem Int Ed Engl 32:666
5. Marr JJ (1996) Drug Discovery Today 1:94
6. Ellington AD, Szostak JW (1990) Nature 346:818
7. (a) Roush W (1997) Science 276:1192; (b) Rawls RL (1997) Chem Eng News (June 2) 35
8. Takayama KM, Inouye M (1990) Crit Rev Biochem 25:155
9. (a) Szymkowski DE (1996) Drug Discovery Today 1:415; (b) Matteucci MD, Wagner RW (1996) Nature Suppl 384:20
10. See various chapters In: Chadwick DJ, Cardew G (eds) (1997) Oligonucleotides as therapeutic agents, Ciba Found Symp 209, Wiley, Chichester
11. Cooney M, Czernuszewicz G, Postel EH, Flint SJ, Hogan ME (1988) Science 241:456
12. Moser HE, Dervan PB (1987) Science 238:645
13. Le Doan T, Perrouault L, Praseuth D, Habhoub N, Decout JL, Thuong NT, Lhomme J, Hélène C (1987) Nucleic Acids Res 15:7749
14. Grigoriev M, Praseuth M, Guieyesse AL, Robin P, Thuong NT, Hélène C, Harel-Bellan A (1993) Compt Rend Acad Sci Ser III Sci Vie 316:492
15. (a) Collier DA, Thuong NT, Hélène C (1991) J Am Chem Soc 113:1457 and references cited therein; (b) Esudé C, Nguyen CH, Kukreti S, Janin Y, Sun JS, Bisagni E, Garestier T, Hélène C (1998) Proc Natl Acad Sci USA 95:3591
16. (a) Robles J, McLaughlin LW (1997) J Am Chem Soc 119:6014; (b) Szewczyk JW, Baird EE, Dervan PB (1996) J Am Chem Soc 118:6778
17. Nielsen PE, Egholm M, Berg RH, Burchardt O (1991) Science 254:1497
18. Lue J, Bruice TC (1998) J Am Chem Soc 120:1115 and references cited therein
19. (a) Kutyavin IV, Gamper HB, Gall AA, Meyer RB, Jr (1993) J Am Chem Soc 115:9303; (b) Ye X, Kimura K-i, Patel DJ (1993) J Am Chem Soc 115:9325
20. Matteucci M, Lin K-Y, Huang T, Wagner R, Sternbach DD, Mehrotra M, Besterman JM (1997) J Am Chem Soc 119:6939
21. Soyfer VN, Potaman VN (1996) Triple-helical nucleic acids. Springer, Berlin Heidelberg New York
22. Sun JS, Mergny JL, Lavery R, Montenay-Garestier T, Hélène C (1991) J Biomol Struct Dynamics 9:411
23. Ji J, Hogan ME, Gao X (1996) Structure 4:425
24. (a) Liu K, Sasisekharan V, Miles HT, Raghunathan G (1996) Biopolymers 39:573; (b) Betts L, Josey JA, Veal JM, Jordan SR (1995) Science 270:1838; (c) Arnott S, Bond PJ, Selsing E, Smith PJC (1976) Nucleic Acids Res 3:2459
25. Srinivasan AR, Olson WK (1998) J Am Chem Soc 120:484
26. Doronina SO, Behr JP (1997) Chem Soc Rev 63
27. Birikh KR, Heaton PA, Eckstein F (1997) Eur J Biochem 3:429
28. Morgan RA, Anderson WF (1993) Annu Rev Biochem 62:191

29. (a) Bishop JS, Guy-Caffey JK, Ojwang JO, Smith SR, Hogan ME, Cossum PA, Rando RF, Chaudhary N (1996) J Biol Chem 271:5698; (b) Gold L, Poliski B, Uhlenbeck O, Yarus M (1995) Annu Rev Biochem 64:763; (c) Mishra RK, LeTinévez R, Toulmé JJ (1996) Proc Natl Acad Sci USA 93:10679

30. (a) Macaya RF, Schultze P, Smith FW, Roe JA, Feigon J (1993) Proc Natl Acad Sci USA 90:3745; (b) Wang KY, Krawczyk SH, Bischofberger N, Swaminathan S, Bolton PH (1993) Biochemistry 32:11285

31. Fire A, Xu S, Montgomery MK, Kostas SA, Driver SE, Mello CC (1998) Nature 391:806

32. Miller PS, Agris CH, Aurelian L, Blake KP, Glave SA, Lin S-B, Murakami A, Reddy PM, Smith CC, Spitz SA, Ts'O POP (1987) Matagen: a family of sequence specific oligonucleoside methylphosphonates. In: Chagas C, Pullman B (eds) Molecular mechanisms of carcinogenic and antitumor activity, Adenine Press, Schenectady, p 169

33. Moses AC, Shepartz A (1997) J Am Chem Soc 119:11591

34. De Mesmaeker A, Häner R, Martin P, Moser HE (1995) Acc Chem Res 28:366

35. Cohen JS (1993) Phosphorothioate oligodeoxynucleotides. In: Crooke ST, Lebleu B (eds) Antisense research, applications, CRC Press, Boca Raton, p 205

36. Hyrup B, Nielsen PE (1996) Bioorg Med Chem 4:5

37. Wang S, Kool ET (1994) J Am Chem Soc 116:8857

38. Giovannangeli C, Thuong NT, Hélène C (1993) Proc Natl Acad Sci USA 90:10013

39. Boutorine AS, Brault D, Takasugi M, Delgado O, Hélène C (1996) J Am Chem Soc 118:9469 and references cited therein

40. (a) Kutyavin IV, Gamper HB, Gall AA, Meyer RB, Jr (1993) J Am Chem Soc 115:9303; (b) Maruenda H, Tomasz M (1996) Bioconjugate Chem 7:541 and references cited therein

41. Sabat M, Lippert B (1996) Met Ions Biol Syst 33:143 and references cited therein

42. Durland RH, Kessler DJ, Gunnell S, Duvic M, Pettitt BM, Hogan ME (1991) Biochemistry 30:9246

43. Kohwi Y, Kohwi-Shigematsu (1988) Proc Natl Acad Sci USA 85:3781

44. Laughlan G, Murchie AIH, Norman DG, Moore MH, Moody PCE, Lilley DMJ, Luisi B (1994) Science 265:520

45. Cheong C, Moore PB (1992) Biochemistry 31:8406

46. Kang C, Zhang X, Ratliff R, Moyzis R, Rich A (1992) Nature 356:126

47. Lilley DMJ (1990) The structure of the helical four-way junction in DNA, and its role in genetic recombination. In: Eckstein F, Lilley DMJ (eds) Nucleic acids and molecular biology, vol 4. Springer, Berlin Heidelberg New York, p 55

48. Pecinka P, Huertas D, Azorin F, Palecek E (1995) J Biomol Struct Dynamics 13:29

49. (a)Bernués J, Beltrán R, Casasnovas JM, Azorin F (1989) The EMBO J 8:2087; (b) Beltrán R, Martinez-Balbás A, Bernués J, Bowater R, Azorin F (1993) J Mol Biol 230:966; (c) Bernués J, Azorin F (1995) Triple-stranded DNA. In: Eckstein F, Lilley DMJ (eds) Nucleic acids and molecular biology, vol 9. Springer, Berlin Heidelberg New York, p 1 and references cited therein

50. Potaman VN, Soyfer VN (1994) J Biomol Struct Dynamics 11:1035

51. (a) Cowan JA (1997) J Biol Inorg Chem 2:168 and references cited therein; (b) Basile LA, Barton JK (1989) Met Ions Biol Syst 25:31

52. (a) Feig AL, Scott WG, Uhlenbeck OC (1998) Science 279:81; (b) Ohmichi T, Sugimoto N (1997) Biochemistry 36:3514; (c) Zhou DM, Kumar PKR, Zhang LH, Taira K (1996) J Am Chem Soc 118:8969; (d) Pyle AM (1996) Met Ions Biol Syst 32:479; (e) Piccirilli JA, Vyle JS, Caruthers MH, Cech TR (1993) Nature 361:85; (f) Kazakov S, Altman S (1992) Proc Natl Acad Sci USA 89:7939

53. (a) Scott WG, Murray JB, Arnold JRP, Stoddard BL, Klug A (1996) Science 274:2065; (b) Pley HW, Flaherty KM, McKay DB (1994) Nature 372:68

54. (a) Strobel SA, Dervan PB (1990) Science 249:73; (b) Strobel SA, Doucette-Stamm LA, Riba L, Housman DE, Dervan PB (1991) Science 254:1639

55. (a) Chen CB, Sigman DS (1988) J Am Chem Soc 110:6570; (b) Sigman DS, Mazumder A, Perrin DM (1993) Chem Rev 93:2295
56. Pitié M, Meunier B (1996) J Biol Inorg Chem 1:239 and references cited therein
57. Tsubouchi A, Bruice TC (1995) J Am Chem Soc 117:7399
58. See e.g.: (a) Bashkin JK, Frolova EI, Sampath US (1994) J Am Chem Soc 116:5981; (b) Magda D, Miller RA, Sessler JL, Iverson BL (1994) J Am Chem Cos 116:7439; (c) Mutsumura K, Endo M, Komiyama M (1994) J Chem Soc Chem Commun 2019; (d) Hall J, Hüsken D, Pieles U, Moser HE, Häner R (1994) Chem Biol 1:185
59. Magda D, Wright M, Crofts S, Lin A, Sessler JL (1997) J Am Chem Soc 119:6947
60. (a) Corey DR, Pei D, Schultz PG (1989) J Am Chem Soc 111:8523; (b) Kanaya S, Nakai C, Konishi A, Inoue H, Ohtsuka E, Ikehara M (1992) J Biol Chem 267:8492
61. Takeda N, Imai T, Irisawa M, Sumaoka J, Yashiro M, Shigekawa H, Komiyama M (1996) Chem Lett 599
62. (a) Sundquist WI, Lippard SJ (1990) Coord Chem Rev 100:293; (b) Bloemink MJ, Reedijk J (1996) Met Ions Biol Syst 32:641
63. (a) Augé P, Kozelka J (1997) Transition Met Chem 22:91 and references cited therein; (b) Lippert B (1996) Met Ions Biol Syst 33:105; (c) Lippert B (1989) Prog Inorg Chem 37:1
64. (a) Takahara PM, Rosenzweig AC, Frederick CA, Lippard SJ (1995) Nature 377:649; (b) Takahara PM, Frederick CA, Lippard SJ (1996) J Am Chem Soc 118:12309
65. Vlassov VV, Gorn VV, Ivanova EM, Kazakov SA, Mamaev SV (1983) FEBS Lett 162:286
66. Chu BCF, Orgel LE (1990) Nucleic Acids Res 18:5163
67. (a) Strothcamp KG, Lippard SJ (1976) Proc Natl Acad Sci USA 73:2536; (b) Szalda DJ, Eckstein F, Sternbach H, Lippard SJ (1979) J Inorg Biochem 11:279
68. (a) Huang H, Zhu L, Reid BR, Drobny GP, Hopkins PB (1995) Science 270:1842; (b) Paquet F, Pérez C, Leng M, Lancelot G, Malinge JM (1996) J Biomol Struct Dynamics 14:67
69. Brabec V, Leng M (1993) Proc Natl Acad Sci USA 90:5345
70. Krizanovic O, Sabat M, Beyerle-Pfnür R, Lippert B (1993) J Am Chem Soc 115:5538
71. Metzger S, Erxleben A, Lippert B (1997) J Biol Inorg Chem 2:256
72. Metzger S, Britten JF, Erxleben A, Lock CJL, Albinati A, Lippert B (1999) (submitted)
73. Sigel RKO, Freisinger E, Metzger S, Lippert B (1998) J Am Soc 120:12000
74. Metzger S, Lippert B (1996) J Am Chem Soc 118:12467
75. Brabec V, Sip M, Leng M (1993) Biochemistry 32:11676
76. Paquet F, Boudvillain M, Lancelot G, Leng M, unpublished results
77. Zamora F, Witkowski H, Freisinger E, Müller J, Thormann B, Albinati A, Lippert B (1998) J Chem Soc Dalton Trans (in press)
78. Dieter-Wurm I, Sabat M, Lippert B (1992) J Am Chem Soc 114:357
79. Berghoff U, Schmidt K, Janik M, Schröder G, Lippert B (1998) Inorg Chim Acta 269:135
80. Schmidt K, Janik M, Lippert B, unpublished results
81. Schliepe J, Berghoff U, Lippert B, Cech D (1996) Angew Chem Int Ed Engl 35:646
82. Lepre CA, Lippard SJ (1990) Interaction of platinum antitumor compounds with DNA. In: Eckstein F, Lilley DMJ (eds) Nucleic acids and molecular biology, vol 4. Springer, Berlin Heidelberg New York, p 9
83. Reedijk J (1987) Pure Appl Chem 59:181
84. Comess KM, Costello CE, Lippard SJ (1990) Biochemistry 29:2102
85. Dalbiès R., Boudvillain M, Leng M (1995) Nucleic Acids Res 23:949
86. Dalbiès R, Payet D, Leng M (1994) Proc Natl Acad Sci USA 91:8147
87. Boudvillain M, Guérin M, Dalbiès R, Saison-Behmoaras T, Leng M (1997) Biochemistry 36:2925
88. Colombier C, Boudvillain M, Leng M (1997) Antisense and Nucleic Acid Drug Development 7:397
89. Boudvillain M, Dalbiès R, Aussourd C, Leng M (1995) Nucleic Acids Res 23:2381

90. Prévost C, Boudvillain M, Beudaert P, Leng M, Lavery R, Vovelle F (1997) J Biomol Struct Dynamics 14:703
91. Gaucheron F, Malinge JM, Blacker AJ, Lehn JM, Leng M (1991) Proc Natl Acad Sci USA 88:3516
92. Hollis LS, Amundsen AR, Stern EW (1989) J Med Chem 32:128
93. Payet D, Leng M (1994) DNA, cis-platinum and heterocyclic amines: catalytic activity of the DNA double helix. In: Sarma RH, Sarma MH (eds) Structural biology: the state of the art, vol 2. Adenine Press Guilderland, New York, p 325
94. Payet D, Gaucheron F, Sip M, Leng M (1993) Nucleic Acids Res 21:5846
95. Lemaire MA, Schwartz A, Rahmouni AR, Leng M (1991) Proc Natl Acad Sci USA 88:1982
96. Colombier C, Lippert B, Leng M (1996) Nucleic Acids Res 24:4519
97. Bernal-Mendez E, Sun JS, Gonzalez-Vilchez G, Leng M (1998) New J Chem 1479
98. (a) Van Meervelt L, Vlieghe D, Dautant A, Gallois B, Précigoux G, Kennard O (1995) Nature 374:742; (b) Vlieghe D, Van Meervelt L, Dautant A, Gallois B, Précigoux G, Kennard O (1996) Science 273:1702
99. Lippert B, results to be published
100. Johansson HE, Blesham GJ, Sproat BS, Hentze MW (1994) Nucleic Acids Res 22:4591
101. Manchanda R, Dunham SU, Lippard SJ (1996) J Am Chem Soc 118:5144

Sulfoxide Ruthenium Complexes: Non-Toxic Tools for the Selective Treatment of Solid Tumour Metastases

Gianni Sava[1,3], Enzo Alessio[2], Alberta Bergamo[3], Giovanni Mestroni[2]

[1]Department of Biomedical Sciences, [2]Department of Chemical Sciences, University of Trieste, and Fnd. Callerio, [3]Institutes of Biological Research, via A. Fleming 22–31, I-34127-Trieste, Italy
E-mail: gsava@fc.univ.trieste.it

The main characteristics that make ruthenium-sulfoxide complexes different from other anti-tumour drugs are the selectivity for solid tumour metastases and the lack of significant host toxicity at pharmacologically active dosages. The complexes studied so far are characterised by the presence of dimethylsulfoxide ligands and, those which share the above properties, also by N-donor ligands on ruthenium (III) species. NAMI-A, [ImH][trans-RuCl$_4$(DMSO)Im], is the best result of such investigation. It derives from NAMI by replacing Na$^+$ with ImH$^+$, acquiring more stability, reproducibility of elemental analysis and better pharmacological properties. NAMI-A is a ruthenium compound which may easily undergo reduction in vivo, with rapid loss of a chlorine ligand. It binds DNA and produces effects similar to those of cisplatin but, unlike cisplatin, only at relatively high concentrations of [trans-RuCl$_4$(DMSO)Im]$^-$, and binds also transferrin, an event which is supposed to target the compound to cancer cells. NAMI-A causes a marked reduction of lung metastases in mice carrying solid metastasising tumours. This effect is increased by surgery of primary tumour and is pronounced also when NAMI-A is given to mice with advanced metastases. The antimetastatic effect of NAMI-A is not related to a direct cytotoxicity of the compound for cancer cells. Rather, it seems that it modifies the relationships between cancer cells and host tissues in favour of a reduced capacity of cancer growth.

Keywords. Ruthenium, Metastasis, Solid tumours, Treatment

1
Introduction

1.1
Foreword

Metastases of solid tumours represent the main reason for the failure of cancer therapy. In fact, while surgery and/or radiotherapy may successfully cure the primary lesions, many human tumours develop distant metastases which, independent of their diagnosis at the beginning of therapy, lead invariably to death. Because of their almost always scattered location, drug therapy appears to be the best choice for their treatment. In fact, theoretically, drugs can distribute in the body and reach metastatic lesions in any possible location. However, at present, in such an approach, the weak ring of the chain is represented by the drugs available.

These drugs, often the result of the so-called 'serendipitous discoveries', are mainly characterised by the fact that they interact with cell division and growth by a cytotoxic mechanism often related to a direct interaction with DNA or DNA related mechanisms of cell division, are only scarcely tumour-cell specific and have dose-limiting toxicity due to bone marrow and immune-response impairment. Indeed, it should be stressed that the most important transversal link which is commonly shared by these drugs is the fact that they have been preclinically studied using in vitro systems or in vivo models with the principal aim of inhibiting primary tumour growth independent of the test system. Conversely, the last two decades have contributed highly to the demonstration of the peculiarity of solid tumour metastases in regard to the differences from the primary tumours from which they arise and, in particular, the different chemical sensitivity to the cytotoxic agents available compared to their primary counterparts. Therefore, specific anti-metastatic drugs are being actively sought.

The square planar platinum(II) complex, cis-$[PtCl_2(NH_3)_2]$ (cis-DDP, cisplatin), is the world's leading anti-tumour drug for the chemotherapy of human cancer [1]. The complex has a high cure rate against testicular carcinomas, and increases significantly the life expectancy of patients with ovarian tumours, head and neck cancers, bladder tumours and osteosarcomas. However, several tumours may have spontaneous (or acquired) resistance to cisplatin. For example, it shows only minor or insufficient activity against a number of malignancies with high social incidence, such as lung carcinomas and adenocarcinomas of the colon and rectum, that are responsible for 30% of cancer mortality.

The efficacy of cisplatin is further limited by its rather strong toxic side effects. Second and third generation platinum drugs, which are currently being brought into clinical use, even though they have some improved characteristics compared to the parent compound in terms of reduced host toxicity, do not seem able to broaden the spectrum of action of cisplatin significantly. As for the mechanism of action, it is now accepted that the primary target of platinum drugs is DNA, to which they bind covalently, most frequently to neighbouring guanine-N7 sites. Chelation of platinum induces distortions of the double helix that affect both replication and transcription and ultimately lead to cell death [2].

These findings have, for a long time, stimulated investigations into the field of non-platinum metal anti-tumour drugs [3]. Non-platinum active compounds are likely to have a mechanism of action, biodistribution and toxicity different from those of platinum drugs and might therefore be active against human malignancies that are resistant, or have acquired resistance, to them. Many different classes of coordination compounds and organometallic derivatives have been screened on model tumours and some promising results have been obtained with derivatives of different metals, e.g. tin [4], gold [5], titanium [6] and, in particular, ruthenium [7].

1.2
Ruthenium Complexes in Biological Systems

Ruthenium complexes are found mainly as ruthenium(II) and ruthenium(III) in aqueous solution; in both oxidation states the metal is hexacoordinated, with a roughly octahedral geometry. Like platinum(II), ruthenium ions have a high affinity for nitrogen and sulfur donor ligands. Ruthenium coordination complexes with diverse coordination environments have in the last two decades achieved several promising results in the biological field and not only as anti-tumour agents, as schematically summarised below:

1980 Anti-tumour activity of fac-[RuCl$_3$(NH$_3$)$_3$] and cis-[Ru(NH$_3$)$_4$Cl$_2$]Cl in mice [8]

1984 Anti-tumour activity of cis-[RuCl$_2$(Me$_2$SO)$_4$] in mice [9]

1986 Anti-tumour activity of [LH][$trans$-RuCl$_4$L$_2$] compounds (L=heterocyclic nitrogen ligand) against a platinum-resistant colorectal tumour in rats [10]

1988 Anti-tumour and anti-metastatic activity of $trans$-[RuCl$_2$(Me$_2$SO)$_4$] in mice [11]

1992 Anti-metastatic activity of Na[$trans$-RuCl$_4$(Me$_2$SO)(L)] compounds against spontaneous lung metastases in mice [12]

1995 Ru(III)-polyaminocarboxylates as NO scavengers for the treatment of septic shock[13]

1995 Anti-tumour activity in mice and in vitro cytotoxicity of *mer*-[RuCl₃(terpy)] [14]

1995 Anti-tumour activity in mice and in vitro cytotoxicity of [H][Ru^{III} (H$_2$pdta)Cl$_2$]·4H$_2$O (pdta=1,2-propylenediaminetetraacetate) [15].

1996 Immunosuppressive activity of Ru(III)-N(heterocyclic) compounds at nanomolar concentrations [16]

Since ruthenium(III) complexes are normally more inert than the corresponding ruthenium(II) derivatives, an *"activation by reduction"* mechanism has been proposed to explain the anti-tumour activity found in simple chloro-ammino ruthenium derivatives, such as *cis*-[Ru(NH₃)₄Cl₂]Cl and *fac*-Ru(NH₃)₃ Cl₃, that are considerably inert towards ligand loss opening up coordination positions [7,8]. According to this hypothesis, the inert, and therefore inactive, ruthenium(III) complexes are considered as pro-drugs that can be activated by an in situ reduction to the corresponding less inert ruthenium(II) species. These should be able to form covalent bonds with biological targets after relatively rapid dissociation of some ligands. Ruthenium(III) species might be expected to be reduced more easily in tumour masses which are generally considered as reducing, hypoxic environments compared to surrounding, more aerated tissues. Therefore larger amounts of reactive ruthenium(II) species might be generated in tumour tissues, compared to healthy ones, thus promoting accumulation of ruthenium and ultimately leading to a selective cytotoxicity against solid tumours. Ruthenium(III) species with biologically accessible reduction potentials are obviously more likely to be involved in such a mechanism.

1.3
Anti-Tumour Active Ruthenium Sulfoxide Complexes

In the last few years we have been working towards the development of ruthenium(II) and ruthenium(III) complexes endowed with anti-tumour properties [17]. The presence of coordinated dimethylsulfoxide (Me₂SO) has been a hallmark in our complexes. Dimethylsulfoxide is an ambidentate ligand that can bind to a metal centre either through the sulfur atom (Me₂SO) or through the oxygen atom (Me₂SO). The choice of Me₂SO as ligand was originally motivated by the following considerations:

(i) Me₂SO is a polar molecule and is known to easily cross cell membranes. Therefore, when bound to a metal centre, Me₂SO might give appreciable water solubility to the complex and improve its capability of crossing biological membranes.

(ii) When bound through sulfur, Me₂SO also has a rather strong *trans* effect, which promotes the formation of free coordination sites on the metal centre. Moreover, Me₂SO is a mild π-acceptor and stabilises metals in low oxidation states. This is particularly important in view of the possibility that inert ruthenium(III) complexes might undergo reduction in vivo, thus generating more labile, and therefore more active, ruthenium(II) species.

We initially investigated the anti-tumour activity and DNA binding pattern of simple ruthenium(II) compounds, such as *cis*- and *trans*-[RuCl$_2$(Me$_2$SO)$_4$] [18]. The anti-tumour testing of *cis*-[RuCl$_2$(Me$_2$SO)$_4$], compared to cisplatin, showed a particular effectiveness on Lewis lung carcinoma, MCa mammary carcinoma and B16 melanoma. The use of *cis*-[RuCl$_2$(Me$_2$SO)$_4$] on these tumours appeared to be advantageous compared to cisplatin since, unlike with cisplatin, the anti-tumour effects were observed at doses with reduced host toxicity, indicated by the absence of significant haematological toxicity and toxicity for normal proliferating tissues [9]. However, it must be stressed that such effects were obtained at doses equal to or greater than 600 mg/kg/day given for up to 14 consecutive days, a dosage much greater than that typically used for cisplatin also in molar terms (1238 vs 1.7 μmols, respectively).

1.4
Discovery of the Selective Anti-Metastatic Action of Ruthenium Sulfoxide Complexes

Later experiments showed that the geometrical isomer *trans*-[RuCl$_2$(Me$_2$SO)$_4$], given for 10 consecutive days after surgical removal of the primary tumour at 76 μmol/kg/day, caused a significant prolongation of the life expectancy of animals with Lewis lung carcinoma, suggesting an interesting therapeutic potential when combined with surgical amputation of the primary tumour [19]. Provided that the survival time of the tumour bearing mice in these experimental conditions depends solely on the growth of lung metastases, this effect of *trans*-[RuCl$_2$(Me$_2$SO)$_4$] has to be ascribed to the reduction of metastasis growth. This effect constitutes the first example of anti-metastatic action for ruthenium complexes.

More recently, we have developed two classes of anionic and neutral ruthenium(III) compounds of the general formula Na[*trans*-RuCl$_4$(Me$_2$SO)(L)] and *mer,cis*-[RuCl$_3$(Me$_2$SO)$_2$(L)], respectively, where L is either ammonia or an heterocyclic nitrogen donor ligand [20]. This research was also stimulated by the interesting results obtained by Keppler et al. on murine tumours of different origin with the new class of ruthenium(III) complexes [LH][*trans*-RuCl$_4$L$_2$] (L= heterocyclic nitrogen ligand) [21].

We found that the neutral ammino complex *mer*-[RuCl$_3$(Me$_2$SO)$_2$NH$_3$] was as effective as cisplatin on primary tumour growth of Lewis lung carcinoma bearing mice, but was much more effective than cisplatin on the prolongation of host survival time. Conversely, on MCa mammary carcinoma the effects on primary tumours were lower than those of cisplatin but were equivalent on host survival time. Qualitatively the anti-tumour action of *mer*-[RuCl$_3$(Me$_2$SO)$_2$NH$_3$] was different from that of cisplatin and, similar to ruthenium(II) complexes tested previously, seemed to be particularly effective on tumours localised in the lungs [22]. Other members of the *mer,cis*-[RuCl$_3$(Me$_2$SO)$_2$(L)] class of neutral complexes had, however, a limited solubility in water and therefore we investigated in more detail the anionic complexes Na[*trans*-RuCl$_4$(Me$_2$SO)(L)]. Preliminary screenings on animal tumour models gave evidence for the selective anti-meta-

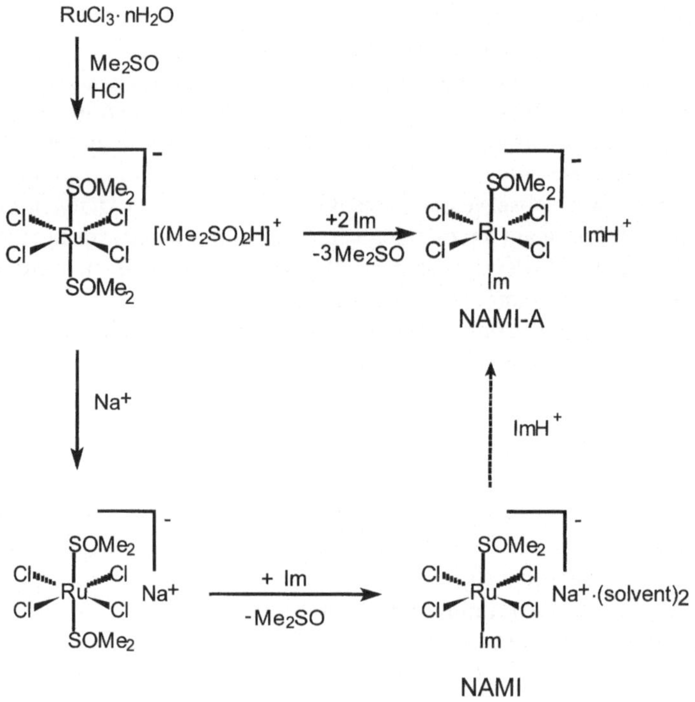

Scheme 1. Synthetic pathways leading to NAMI and NAMI-A

static properties of the imidazole (Im) derivative Na*trans*-[RuCl$_4$(Me$_2$SO)(Im)] (NAMI) [12], which was thus selected for further investigation. NAMI is obtained from hydrated RuCl$_3$ in three steps (Scheme 1).

Commercially available ruthenium trichloride is first warmed in a mixture of dimethyl sulfoxide and aqueous HCl to give [(Me$_2$SO)$_2$H][*trans*-RuCl$_4$(Me$_2$SO)$_2$], which is then transformed into the corresponding sodium salt. The latter is finally treated with imidazole in a dimethyl sulfoxide/acetone mixture to give the product, usually formulated with two Me$_2$SO molecules of crystallisation, i.e. as Na*trans*-[RuCl$_4$(Me$_2$SO)(Im)]·2Me$_2$SO.

2
Selective Treatment of Solid Tumour Metastases

Ruthenium(III) complexes, characterised by sulfoxide and N-donor ligands, are the most recent group of compounds for which anti-metastatic properties have been repeatedly reported. One of them, NAMI, might represent a suitable model compound for anti-metastatic activity.

2.1
NAMI: The Most Studied Anti-Metastatic Ruthenium Compound

As stated previously, the anionic complex Na$trans$-[RuCl$_4$(Me$_2$SO)(Im)] (NA-MI), appeared to be the most promising among those tested in that:

(1) it is more soluble in water compared to the neutral species mer-[RuCl$_3$(Me$_2$SO)$_2$NH$_3$];
(2) similar to cisplatin, though at a lower level, it reduces tumour growth in its primary site (~40% vs controls) in each tumour model used (Lewis lung carcinoma, B16 melanoma, MCa mammary carcinoma);
(3) unlike cisplatin, it increases the life span of the host independent of the tumour treated; and
(4) it reduces lung metastasis formation even when the effects on the primary tumour are completely negligible [23].

NAMI is capable of reducing primary tumour growth and of prolonging the survival time with different schedules of administration. However, better effects were obtained (a) with treatments started soon after tumour implantation, (b) with daily administration rather with treatments at 4-d intervals, and (c) using relatively low doses (22–66 mg/kg/day for up to 12 days). The prolongation of survival time, greater than that obtained with cisplatin in parallel experiments, is not simply related to the effects on primary tumour growth, which is always less than that of cisplatin [12]. More notable is the reduction of lung metastasis formation, both when they formed spontaneously from intramuscular (i.m.) tumour implants and when they were obtained by intravenous (i.v.) injection of tumour cells.

A detailed comparison between the ruthenium(II) and ruthenium(III) sulfoxide complexes examined by us, independent of their chemical behaviour, seems to highlight their anti-metastatic properties rather than a merely cytotoxic cell killing like that caused by cisplatin [24].

The greater reduction of lung metastases of Lewis lung carcinoma and of MCa mammary carcinoma as compared to the effects on primary tumour growth following i.v. treatment with NAMI were also evidenced by vivo vivo bioassays. These studies showed that when tumour cells of the primary tumour of mice treated in vivo with NAMI were implanted into healthy syngenic animals, tumour take up and growth in the site of the primary implant was similar to controls but all animals showed a marked reduction of metastatic ability [25]. It thus seems that lung metastases of these tumours are invariably the elective target for the pharmacological effects of NAMI. Considering that survival time of mice bearing solid metastasising tumours is influenced by both tumour and metastasis growth, the marked anti-metastatic effects caused by NAMI in combination with surgical removal of the primary tumour may account for the prolongation of survival time and for the cure rate [26].

The favourable effect on survival time is also influenced by the lack of significant cytotoxicity for normal tissues such as lung and kidney epithelia, muscle and liver cells, splenocytes and bone marrow [27]. Experimental results showed that the selective interaction with tumour cells in the lungs cannot simply be attributed to a selectively higher localisation of the compound at this site, nor to a modification of the histological structure of the primary tumour itself [27, 28].

By examining a number of intraperitoneal (i.p.)and intravenous (i.v.) administration schedules in mice either with spontaneous lung metastases or after induction of artificial lung colonies, NAMI was further found to behave differently to cisplatin. Unlike cisplatin, which is equally active on primary tumour growth and lung colonies, the effects of NAMI were always more pronounced on spontaneous metastases than on primary tumour growth. Concerning lung colonies (i.e. metastases obtained by i.v. injection of tumour cells), NAMI showed an activity comparable to that of cisplatin on the reduction of the number of colonies but was much less effective than cisplatin on the survival time of the same animals [26]. Considering that lung colonies are much more similar to primary tumours rather than to true spontaneous metastases, these results show that NAMI, unlike cisplatin, discriminates between spontaneous metastases and lung colonies and is less effective on the latter as it is scarcely effective on primary tumour growth.

In combination with surgery, NAMI prevented metastasis formation and inhibited the growth of those already formed (Fig. 1). This effect, although dependent on the dose used, was not associated with any residual effect on primary tumour cells after treatment discontinuation, whereas it tended to reduce the metastatic ability of the same tumour. This observation stresses the particular propensity of NAMI to attack metastatic cells rather than other tumour cell clones [26].

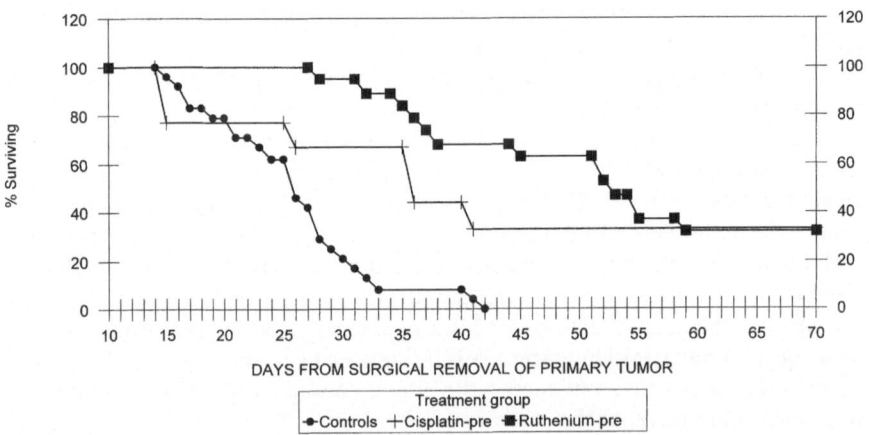

Fig. 1. Prolongation of life expectancy of mice with MCa mammary carcinoma treated i.v. with 60 mg/kg/day NAMI or 2.7 mg/kg/day cisplatin prior to surgical ablation of primary tumour

2.1.1
Cytotoxic Effects In Vivo and In Vitro

The relevance of cytotoxicity in the anti-metastasis effect of NAMI was tested in comparison with other analogues of the series Na$trans$-[RuCl$_4$(Me$_2$SO)(L)] and Na$trans$-[RuCl$_4$(TMSO)(L)] (TMSO=tetramethylenesulfoxide). These studies demonstrated that in vitro cytotoxicity is present only at high concentrations ($>10^{-4}$ M), depends on lipophilicity and is markedly affected by the presence of serum and plasma proteins in culture medium [25]. The most lipophilic complex Na$trans$-[RuCl$_4$(TMSO)Iq] (TEQU; Iq=isoquinoline) causes the same dose-dependent cytotoxicity and DNA fragmentation shown by cisplatin, while NAMI is virtually devoid of any detectable effect [26]. The study of in vivo cytotoxicity was carried out with TLX5 lymphoma cells injected i.p. in CBA mice. NAMI and TEQU were given i.p. as single dose or as repeated administrations for seven consecutive days. In all cases we observed a moderate, still significant, increase in survival time in the host with no concomitant significant decay of the number of TLX5 lymphoma cells in the peritoneal cavity of the same animals at the end of the treatment [25]. Additional studies with in vitro treatment of tumour cells followed by in vivo bioassay to evaluate their clonogenic ability and capacity to colonise the brain confirmed the lack of correspondence between an increase in the survival time of the bioassayed animals and a reduction in the number of peritoneal tumour cells, and the direct relationship between the host's survival and reduction of brain involvement by the tumour [25]. The comparison of the effects on in vitro cytotoxicity with in vivo anti-tumour and anti-metastatic action demonstrates that these compounds reduce metastasis formation by a mechanism unrelated to a direct cell cytotoxicity. If on the one hand TEQU, the compound that shows the most potent in vitro cytotoxic effects, is the least effective against metastases then, on the other, NAMI, the compound that better reduces metastasis formation, is rather devoid of cytotoxic effects on tumour cells kept in vitro [25,29].

2.1.2
Dependence of the Anti-Metastatic Effects on Host Interaction

Data reviewed so far do not support the direct cytotoxicity of ruthenium complexes towards tumour cells; rather, there appears to be a role for the host in determining their activity. Concerning solid metastasising tumours, histological analysis of primary and metastatic sites of growth indicates that NAMI causes no evidence of suffering on primary sites even at relatively high i.v. dosages [27]. Conversely, on the same animals, NAMI causes a dramatic reduction in the number of lung metastases which appear to invade a restricted area and only rarely surround blood vessels completely [27]. In relation to host interactions, we have experimental evidence that seem to exclude interactions with host immunity. Although it is well known that metastases are much more susceptible to host immunity than primary tumours, and that some ruthenium complexes have found advantages in their pharmacological activity by the interaction with immune effectors such as cytokines [30], we did not observe any reaction on metastatic nodules that could be ascribed to immune cells. Furthermore, we may also exclude the possibility that NAMI modifies the antigenicity of a cell, via epigenetic mechanisms (chemical xenogenisation [31]), although other metal compounds showed xenogenising properties on MCa mammary carcinoma [32]. Conversely, in mice with MCa mammary carcinoma, in vivo treatment of tumour cells in immuno-suppressed hosts caused a progressive increase in DNA activity and, starting from the fourth transplant generation, a significant susceptibility of lung metastases to a further treatment in intact mice [33].

These data suggest that NAMI may select a tumour cell population which maintains its capacity to metastasise to the lung but with enhanced sensitivity to the anti-metastatic properties of this compound [33]. At least in part, the selective anti-metastasis effect of NAMI may be ascribed to the effect on the invasive potential of tumour cells in the primary tumour site. NAMI, at doses active in reducing lung metastases below 10% of controls, causes a dramatic imbalance of the ratio between the mRNAs of MMP-2 (a matrix metallo proteinase capable of degrading extra-cellular matrix) and of its specific tissue inhibitor TIMP-2. Correspondingly, at an histological level, a pronounced increase of extra-cellular matrix components appears in tumour parenchyma and particularly around tumour blood vessels (Fig. 2), that might suggest an impairment to tumour cell intravasation and metastasis formation [34].

However, under these conditions, no evidence of tumour cell damage in terms of induction of areas of necrosis or of tumour cell degeneration was found histologically. This observation provides evidence in an animal model of the parallel down regulation of both the metastatic potential and the net potential of type-IV collagenolytic activity induced by a non-cytotoxic agent [34].

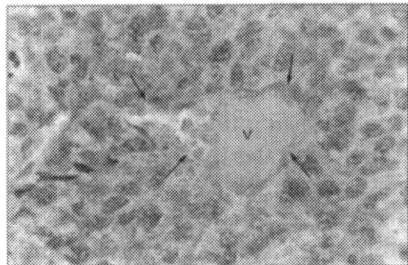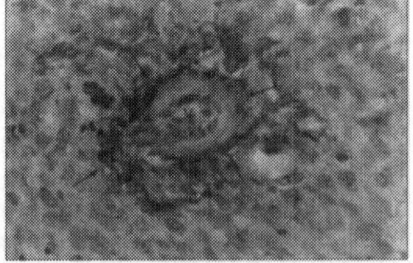

Fig. 2. Histological appearance of a tumour blood vessel inside the primary tumour of MCa mammary carcinoma. *Left hand site*: control; *right hand site*: treated with NAMI. Cajal-Gallego staining. Erythrocytes appear coloured yellow. Green fibers are collagen and connective tissue

2.2
From NAMI to NAMI-A

Despite its reasonable anti-metastatic activity, NAMI is, however, not well suited for pharmacological development for a series of reasons related to the presence of the sodium cation: (i) the complex is hygroscopic; (ii) it has two Me_2SO molecules of crystallisation that can be randomly replaced by water or acetone molecules of the crystallisation solvent, thus inducing a limited analytical reproducibility and a variable molecular weight; (iii) the presence of trace water molecules in the second coordination sphere of NAMI, either from the atmosphere or from the crystallisation process, induces the slow decomposition of the complex in the solid state. Also when stored in a desiccator, samples of NAMI turn green from the original orange colour within a few months, indicating the occurrence of hydrolytic processes.

All these drawbacks were recently solved with the preparation of the corresponding imidazolium salt of NAMI, (ImH)[*trans*-RuCl$_4$(Me$_2$SO)(Im)], named NAMI-A (Fig. 3) [35].

Compared to NAMI, NAMI-A is very stable in the solid state and maintains a good solubility in water. Due to the absence of molecules of crystallisation, NAMI-A has a well-defined molecular weight and its elemental analysis is highly reproducible. A synthetic procedure for the production of very pure NAMI-A in high yield (about 80% with respect to raw ruthenium chloride) has been optimised (Scheme 1).

The new synthesis has one step less than that for NAMI, involving the direct reaction between [(Me$_2$SO)$_2$H]*trans*-[RuCl$_4$(Me$_2$SO)$_2$] and excess imidazole. The same procedure was easily extended to similar compounds of the general formula [LH]*trans*-[RuCl$_4$(Me$_2$SO)(L)]; we have examples with L=NH$_3$, 1-methyl imidazole, pyridine and substituted pyridines [36].

In conclusion NAMI-A, while maintaining all the solution properties essentially equal to those of NAMI, is remarkably more suitable than NAMI for a wide range pharmacological investigation and for possible commercial development as an anti-metastatic drug.

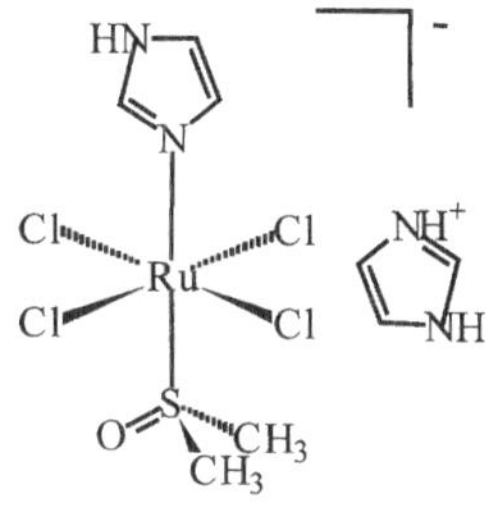

Fig. 3. Schematic drawing of NAMI-A

2.2.1
Chemical Features of trans-[RuCl₄(Me₂SO)(Im)]⁻

The X-ray crystal structure of Na$trans$-[RuCl$_4$(Me$_2$SO)(Im)]·H$_2$O·Me$_2$SO [20] shows that the Ru–S bond distance (2.296(1) Å) is remarkably shorter than the average distance (2.34(1) Å) for Ru–S $trans$ to another Me$_2$SO [37,38]; this shortening was attributed to a decreased competition for π-electrons between the two $trans$-ligands induced by the replacement of the mild π-acceptor Me$_2$SO with imidazole, whose π-acceptor/donor properties are apparently tunable depending on the coordination environment [16]. The Ru–Cl and Ru–N bond lengths in NAMI are very similar to those found in [ImH]$trans$-[RuCl$_4$(Im)$_2$] (ICR) [39], whose anti-tumour properties are also well documented. However, even though NAMI-A is structurally very similar to ICR (formally it can be thought of as deriving from ICR upon replacement of an imidazole with a Me$_2$SO molecule), a comparison of the D$_2$O ^{1}H NMR chemical shifts of the imidazole protons in the two compounds (H2 –5.6, H4 –7.8, H5 –3.5 ppm in NAMI vs H2 –15.8, H4 –21.2, H5 –5.8 ppm in ICR [40], considering NH as position 1) and, above all, of their reduction potentials (+235 mV NAMI vs –275 mV ICR [20]) clearly indicates that the electronic distribution in the two complexes is quite different. The more facile reduction of $trans$-[RuCl$_4$(Me$_2$SO)(Im)]$^-$ compared to $trans$-[RuCl$_4$(Im)$_2$]$^-$ is attributed to the π-acceptor properties of sulfur-bonded dimethylsulfoxide, which greatly stabilises the ruthenium(II) oxidation state.

2.2.2
Chemical Behaviour of trans-[RuCl₄(Me₂SO)(Im)]⁻ in Aqueous Solution

As we have already reported [17], the $trans$-[RuCl$_4$(Me$_2$SO)(Im)]$^-$ anion is rather inert in slightly acidic aqueous solution and chloride hydrolysis occurs slowly. The effect of chloride concentration on the kinetics is marginal. Aquation is accompanied by a drop in pH (from pH 5.0 to 3.86 within 24 h for a 2×10^{-4} M solution), suggesting partial deprotonation of coordinated water molecule(s) with formation of the corresponding hydroxo species. However, under physiological conditions (phosphate buffer pH 7.4, [Cl$^-$]=0.1 M), the complex is considerably more labile. This difference is readily appreciated when comparing the time dependence of the complex main absorption band (390 nm) under different pH conditions (Fig. 4).

The chemical behaviour of $trans$-[RuCl$_4$(Me$_2$SO)(Im)]$^-$ in physiological conditions, as determined by combined NMR and UV/Vis evidence, is summarised in Scheme 2.

Three relatively well-separated steps can be distinguished at 25 °C. The first two correspond to stepwise hydrolysis of chloride groups. While hydrolysis of the first chloride from $trans$-[RuCl$_4$(Me$_2$SO)(Im)]$^-$ is rather fast (approx. 40 min at 25 °C), the second aquation step occurs at a slower rate (approx. 2 h at 25 °C) and is likely to lead to a mixture of cis and $trans$ diaqua species as in the case of ICR [40]. The pK$_a$s of the aqua species have not been determined, but it is very

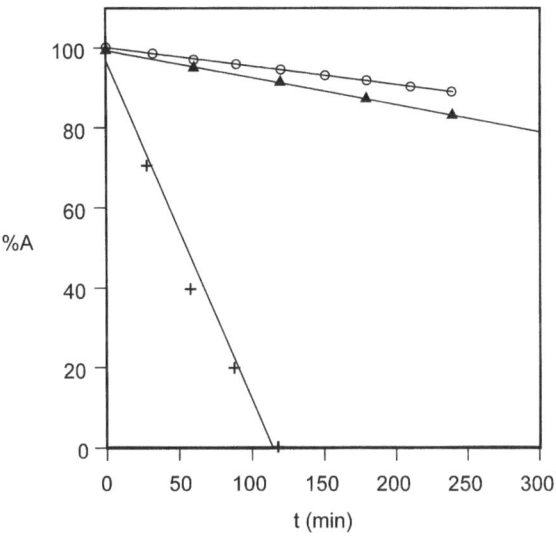

Fig. 4. Time dependence (%) of the main absorption band of NAMI-A at 25 °C in: ○ NaCl 0.9%; ▲ H_2O; + buffer pH=7.4

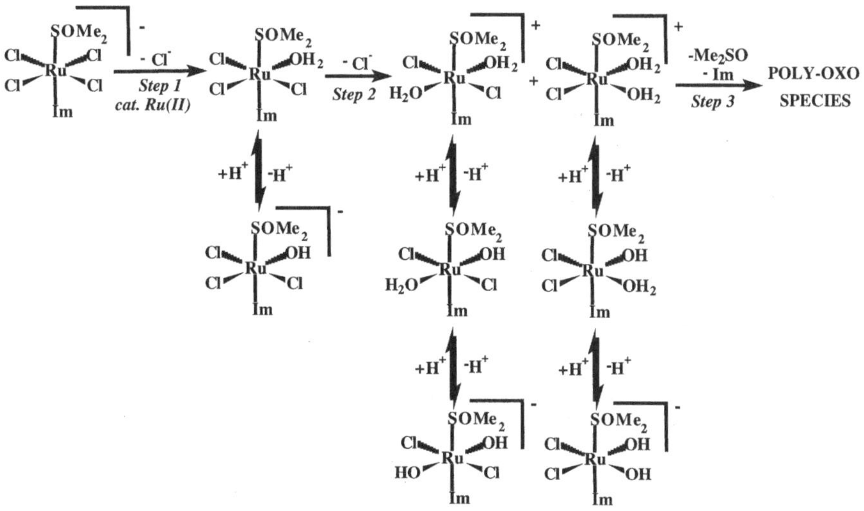

Scheme 2. Chemical behaviour of [*trans*-$RuCl_4(Me_2SO)(Im)$]⁻ under physiological conditions

likely that at physiological pH at least partial deprotonation occurs. For longer observation times, ^{1}H NMR indicates slow partial dissociation of the neutral ligands; this final step is accompanied by a darkening of the solution, suggesting the formation of oxo-bridged polymeric species. We have clear experimental evidence that the rate of the first hydrolytic step is catalysed by ruthenium(II) species. In fact, the kinetic profile of this step is auto-catalytic due to the concomitant formation of ruthenium(II) species via partial self-reduction of *trans*-[RuCl$_4$(Me$_2$SO)(Im)]$^-$; moreover, we found that the reaction rate is considerably enhanced by the addition of traces of biological reductants, such as ascorbic acid or cysteine, which partially reduce *trans*-[RuCl$_4$(Me$_2$SO)(Im)]$^-$ to ruthenium(II) derivatives [17].

However, when stoichiometric amounts of biological reductants are added, e.g. 0.5 equiv of the two-electron reductant ascorbic acid or one equiv of cysteine, *trans*-[RuCl$_4$(Me$_2$SO)(Im)]$^-$ undergoes a very rapid and complete one-electron reduction to ruthenium(II) species. As reported by us in previous studies [17], the reduction is easily monitored by the disappearance of the strong absorption bands in the visible region, characteristic of ruthenium(III) species; in other words, the yellow-orange solution of NAMI-A turns colourless almost immediately upon addition of the reducing agent.

We then investigated by ^{1}H NMR the nature of the ruthenium(II) species generated in the reduction process (Scheme 3). In the first spectrum, taken immediately after addition of the reductant, all resonances are sharp, in accordance with the total reduction of paramagnetic ruthenium(III) to diamagnetic ruthenium(II).

In the first minutes after reduction the prevailing species has one resonance for bound Me$_2$SO (singlet at 3.63 ppm) and three multiplets for bound imidazole (8.53, 7.83, and 7.45 ppm) and we assigned it as *trans*-[RuCl$_4$(Me$_2$SO)(Im)]$^{2-}$. This dianion is then rather rapidly replaced by another species with bound Me$_2$SO (singlet at 3.60 ppm) and imidazole (8.53, 7.79, and 7.51 ppm), which must therefore derive from the first one upon dissociation of one chloride and is formulated as the mono-anion *mer*-[RuCl$_3$(Me$_2$SO)(Im)(H$_2$O)]$^-$. This species reaches a maximum concentration after approximately 1 h (Fig. 5)and then decreases slowly in favour of a number of minor products (see below).

According to this evidence the reduction process of *trans*-[RuCl$_4$(Me$_2$SO) (Im)]$^-$ is considerably faster than the following substitution reactions and can therefore be rationalised as occurring by an outer-sphere mechanism.

During the first hour after reduction no free Me$_2$SO was detected and the free-to-bound imidazole ratio remained reasonably constant at 1, indicating that both neutral ligands are strongly bound to ruthenium(II).

Due to the relative inertness of *mer*-[RuCl$_3$(Me$_2$SO)(Im)(H$_2$O)]$^-$, an NMR titration was performed in order to establish the pK_a of coordinated water. However, in the pH range from 3 to 9, within which the pK_a of water molecules coordinated to ruthenium(II) is generally believed to fall [40], no significant change in the chemical shifts of coordinated imidazole and Me$_2$SO was observed. This result, rather than suggesting that deprotonation of coordinated water does not

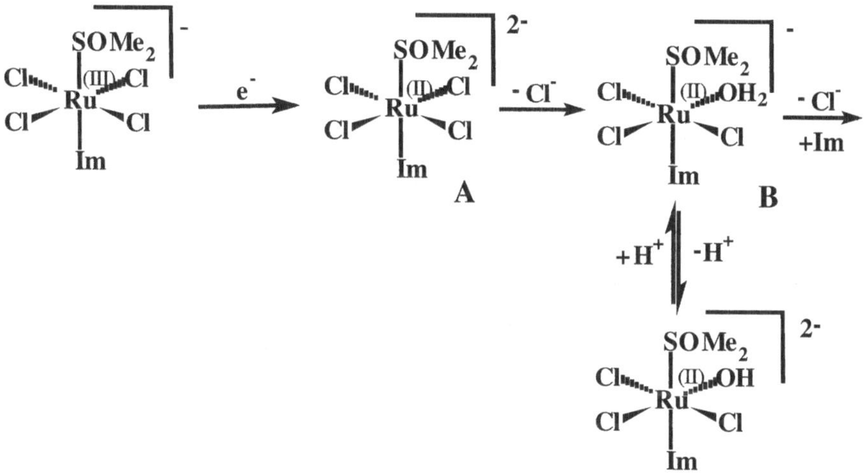

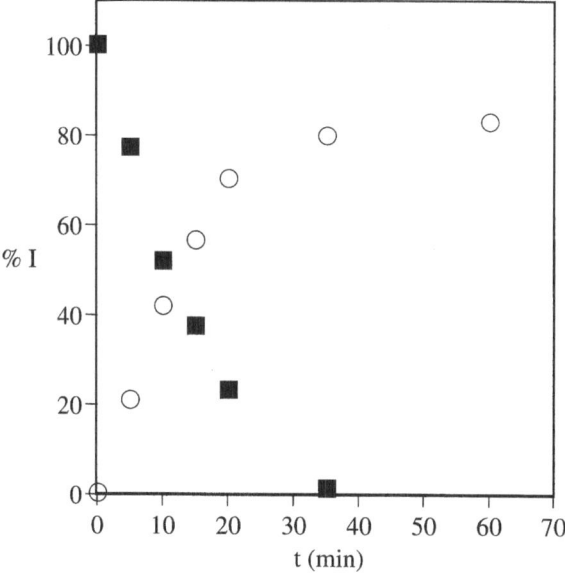

Scheme 3. Chemical behaviour of *trans*-[RuCl$_4$(Me$_2$SO)(Im)]$^-$ under physiological conditions after addition of a slight excess of cysteine or ascorbic acid

Fig. 5. Concentration profile of *trans*-[RuCl$_4$(Me$_2$SO)(Im)]$^{2-}$ (■) and *mer*-[RuCl$_3$(Me$_2$SO)(Im)(H$_2$O)]$^-$ (○) under physiological conditions at 25 °C, according to the integration of their ^{1}H NMR signals

occur within this pH range, indicates that the chemical shifts of the protons on coordinated ligands are insensitive to the presence of either H_2O or OH in the coordination sphere of ruthenium.

As stated above, after approximately 1 h of reduction, several new signals of comparable intensity for coordinated imidazole became visible in the NMR spectrum. Due to the low concentration of the new species and to partial overlap of some signals, neither integration nor a COSY spectrum allowed us to establish with precision the number of new species (at least 5) and the number of coordinated imidazole on each of them. However, it seems reasonable to hypothesize that, while two of them might be the *cis*- and *trans*-dichloro isomers produced by further dissociation of a chloride from the mono-anion, the others have very likely more than one imidazole in their coordination sphere. Chloride hydrolysis followed by partial coordination of free imidazole was observed also in the case of ICR [41]. The polyimidazole species might not be relevant from a biological point of view, since they form in a closed system; however, their formation clearly indicates that the ruthenium(II) species derived from the reduction of NAMI-A might bind to N-donor biomolecules in vivo after hydrolysis of the first chloride.

2.2.3
Interaction with DNA

The cytostatic activity of platinum drugs is generally accepted to involve binding to DNA. The platinum–DNA adducts block DNA and RNA synthesis and eventually induce programmed cell death. Thus, despite the low cytotoxic activity generally manifested by ruthenium–Me$_2$SO complexes, investigation of their interactions with DNA is of primary importance and instrumental to the understanding of the mechanism of their action.

Even though NAMI-A has not yet been specifically investigated, NAMI and other compounds of the same class were studied for their interactions with DNA. The sites of interaction of several Na$trans$-[RuCl$_4$(Me$_2$SO)(L)]2Me$_2$SO complexes with double-stranded pBR322 DNA were determined by means of the primer extension footprinting assay [26]. All the investigated complexes have the same pattern of blocking lesions, showing faint stop bands corresponding to nearly every nucleotide, with the exception of thymine, and more intense stop bands corresponding to guanines. Compared to cisplatin, which is known to have a high affinity for neighbouring guanines, NAMI has a markedly lower selectivity. Also the interstrand cross-linking ability of NAMI on calf thymus DNA was assessed by means of the ethidium bromide fluorescence assay. The ruthenium complex was able to form this type of lesion, which are among those critical for the inactivation of DNA replication, in a dose-dependent manner. However, its effect was lower by approximately one order of magnitude compared to that of cisplatin.

In a yet unpublished study, modifications of natural DNA in a cell-free medium by NAMI were investigated by various methods of biochemical analysis or molecular biophysics [42]. These methods include: binding studies by means of flameless atomic absorption spectrophotometry, measurements of melting curves with the aid of absorption spectrophotometry, measurements of CD spectra, interstrand cross-linking assay employing gel electrophoresis under denaturing conditions, mapping of DNA adducts by means of replication and transcription assays and studies of B to Z transition in DNA by measurements of CD spectra. The results indicate that 80–90% of NAMI binds to calf thymus DNA within 24 h at 37 °C; such binding significantly affects the CD spectrum of calf thymus DNA and stabilises DNA in low salt media (0.01 M $NaClO_4$). In addition, the complex unwinds DNA and inhibits the B to Z transition for poly(dG-dC). DNA interstrand cross-linking efficiency of the ruthenium complex was rather low (~1%). On the other hand, it inhibited DNA or RNA synthesis in vitro by DNA or RNA polymerases at the level of DNA adducts. The termination sites appeared preferentially at guanine residues.

These results are consistent with the view that NAMI induces local conformational alterations on DNA. Interestingly, the extent of these effects on the DNA conformation is largest when DNA is modified by $Na[trans\text{-}RuCl_4(Me_2SO)(Im)]$ compared to $(ImH)trans\text{-}[RuCl_4(Im)_2]$.

These extensive investigations have provided experimental support for previous suggestions that the binding of the ruthenium complexes modifies DNA in a way which is different from the modification by anti-tumoural cisplatin. Thus, the results of this work are also consistent with the hypothesis and support the view that ruthenium drugs which bind to DNA in a manner fundamentally different from that of cisplatin can exhibit altered biological properties including the spectrum and intensity of anti-tumour activity.

Other recent and unpublished results on the inhibition of restriction enzymes induced by NAMI on plasmidic DNA [43] show that DNA damage was observed only at relatively high concentrations of NAMI, significantly larger than those at which cisplatin produces similar effects, in good agreement with the known lower cytotoxicity and systemic toxicity of ruthenium compounds compared to platinum compounds. The authors concluded that DNA may represent a realistic target for the action of ruthenium(III) complexes, provided that substantial amounts of ruthenium are able to enter cells and reach nuclear DNA.

2.2.4
Interaction with Transferrin

Transferrin is the principal iron-transport protein in the serum of vertebrates; it is responsible for the transfer of the essential iron(III) ions through the biological fluids from absorption to storage and utilisation sites. The protein, in its apo-form, binds tightly but reversibly two Fe^{3+} ions per molecule; such binding is associated with a major conformational change [44].

The following features of transferrin chemistry suggest the possibility that transferrin binding might target anti-tumour complexes to cancer cell: (i) besides Fe^{3+}, many other metal ions (M^{2+}, M^{3+}, M^{4+}) can bind transferrin with relatively high affinity; (ii) under normal conditions, transferrin is only 30% saturated with iron; (iii) transferrin receptors are present on virtually all dividing cells; and (iv) malignant cells have a higher iron requirement and thus express a greatly increased level of transferrin receptors.

Some conjugates of apotransferrin with ruthenium(III)–Me$_2$SO anti-cancer drugs have been prepared and characterised by spectroscopic techniques [45]. Combined UV/Vis and CD evidence showed that both NAMI and its indazole analogue or, more precisely, their hydrolysis products interact with apotransferrin (at a 2:1 stoichiometry) and that the interaction occurs in concomitance with the second hydrolytic process. According to CD and NMR spectra it can be ruled out that the bound species is a bare ruthenium(III) ion. Instead, they suggest that the heterocycle remains attached to ruthenium(III). Similar results were obtained in the case of ICR and its indazole analogue [46]; the binding of such compounds to human lactoferrin was also investigated by X-ray crystallographic analysis [47].

A recent study by the group of Clarke, which showed that in vitro cytotoxicity of cis-[RuCl$_2$(NH$_3$)$_4$]Cl and ICR was increased by the presence of added apotransferrin [48], has given support to the hypothesis that transferrin might indeed act as the natural carrier of several ruthenium complexes and deliver them to the cells following the iron route.

2.3
NAMI-A: A Potential Ruthenium-Based Drug for Metastasis Treatment

NAMI-A represents the latest product of the combined studies of the chemical reactivity and stability, on the one hand, and the biological effects, on the other, made with NAMI.

The anti-metastatic efficacy of NAMI-A was first compared with that of NAMI. The study was conducted on two solid metastasising tumours in mice, Lewis lung carcinoma and MCa mammary carcinoma of CBA mice. NAMI was used at a dose (44 mg/kg/day) and a treatment schedule (6 consecutive days) already reported to be optimal for treating these tumours. NAMI-A was therefore used at an equimolar dose of 35 mg/kg/day, administered i.p. per injection per day. In both cases, the tumours were treated at advanced stages of growth, starting on day 6 (for Lewis lung carcinoma) and on day 13 (for MCa mammary carcinoma) from primary tumour implantation. The modification of body weight gain during treatment, caused by these dosages, was similar and almost identical to that observed in untreated controls. The lack of effects on body weight gain is indicative of the lack of significant toxicity for the hosts since it means that they fed and grew normally independent of the treatment.

By comparing the data reported in Tables 1 and 2, it appears that NAMI-A reduces the weight of lung metastases with an efficacy which is the same as (Lewis lung carcinoma) or higher than (MCa mammary carcinoma) that of NAMI. Anyhow, the estimated metastatic tumour mass is reduced by 98 and 84%, respectively, on MCa mammary carcinoma and Lewis lung carcinoma. Thus, NAMI-A maintains and even ameliorates the capacity of NAMI to inhibit the growth of lung metastases of solid metastasising tumours, already reported in detail.

Table 1. Effects of NAMI-A and NAMI on Lewis lung carcinoma

Treatment	Body weight[b]	Primary tumour[c]	Lung metastases[a]			
			Number/animal		Weight (mg)/animal	
		mean±S.E.	mean±S.E.	% Var.[d]	mean±S.E.	% Var.[d]
Controls	–	2.66±0.36	25.0±5.7	–	14.2±4.8	–
NAMI	–1.2%	2.41±0.30	10.7±2.3	–57[e]	2.5±0.7	–82[f]
NAMI-A	0.2%	2.92±0.22	8.8±1.3	–65[e]	2.2±0.3	–84[f]

Groups of 7 mice implanted i.m. in the calf of the left hind leg with 10^6 tumour cells on day 0 were given i.p. 44 mg/kg/day NAMI or 35 mg/kg/day NAMI-A on days 6–11. On day 12, the primary tumour was surgically removed and lung metastases were measured on day 21
a Measured after sacrifice of the animals.
b Percentage variation between the beginning and the end of the treatment with respect to the untreated control group.
c Measured at surgical removal of the primary tumour.
d Percentage variation with respect to the group of untreated controls.
e $p<0.05$ statistically different from the controls.
f $p<0.01$ statistically different from the controls.

Table 2. Effects of NAMI-A and NAMI on MCa mammary carcinoma

Treatment	Body weight[b]	Primary tumour[c]	Lung metastases[a]			
			Number/animal		Weight (mg)/animal	
		mean±S.E.	mean±S.E.	% Var.[d]	mean±S.E.	% Var.[d]
Controls	–	3.08±0.27	39.2±8.1	–	31.4±10.2	–
NAMI	+11.7%	3.07±.031	35.2±17.6	–10	8.0±2.5	–74[e]
NAMI-A	+9.0%	2.59±0.33	6.8±2.2	–83[e]	0.5±0.2	–98[e]

Groups of 7 mice implanted i.m. in the calf of the left hind leg with 10^6 tumour cells on day 0 were given i.p. 44 mg/kg/day NAMI or 35 mg/kg/day NAMI-A on days 13–18. On day 19, the primary tumour was surgically removed and lung metastases were measured on day 26.
a Measured after sacrifice of the animals.
b Percentage variation between the beginning and the end of the treatment with respect to the untreated control group.
c Measured at surgical removal of the primary tumour.
d Percentage variation with respect to the group of untreated controls.
e $p<0.01$ statistically different from the controls.

Table 3. Dose-dependence of the effects of NAMI-A on MCa mammary carcinoma

Dose mg/kg/day	Body weight[b]	Primary tumour[c] mean±S.E.	Lung metastases[a]			
			Number/animal		Weight (mg)/animal	
			mean±S.E.	% Var.[d]	mean±S.E.	% Var.[d]
0	–	1856±307	24.4±5.4	–	4.69±1.17	–
25	0.9%	1469±60	21.8±8.9	–11	5.74±2.79	12
35	–3.7%	1248±251	15.8±4.7	–35	3.05±1.23	–35
50	–6.4%	1779±202	10.2±5.9	–58[e]	1.92±1.20	–59[e]

Groups of 7 mice implanted i.m. in the calf of the left hind leg with 10^6 tumour cells on day 0 were given i.p. 25, 35 or 50 mg/kg/day NAMI-A on days 11–16. Lung metastases were measured on day 25
a Measured after sacrifice of the animals.
b Percentage variation between the beginning and the end of the treatment with respect to the untreated control group.
c Measured 24 h after last dosage.
d Percentage variation with respect to the untreated control groups.
e $p<0.05$ statistically different from the controls.

NAMI-A is well tolerated up to a daily dose of 50 mg/kg given for 6 consecutive days (Table 3), and its anti-metastatic action is dose-dependent. However, compared to the experiment shown in Table 2, in mice with MCa mammary carcinoma which do not undergo surgical amputation of the primary tumour 24 h after the last dosage and which are characterised by a relatively low burden of primary tumour at the beginning of treatment, NAMI-A shows a reduced efficacy on lung metastasis formation. The effects reported in Table 3 are of importance in that they stress how NAMI-A causes a dose-dependent reduction of metastasis formation, unrelated to the corresponding effects on primary tumour growth.

The effects of NAMI-A on primary tumour growth and lung metastasis formation in mice bearing MCa mammary carcinoma were further investigated as a function of the stage of tumour growth. Data reported in Table 4 show that NAMI-A is responsible for a weak reduction in primary tumour growth at any treatment schedule used, either when the compound was given during the first five days after tumour transplantation (reduction by 15%) or on days 16–20 (reduction by 12%). Conversely, the effects on lung metastasis seem to depend on the metastasising ability of the tumour: it is strong when NAMI-A is given on days 11–15, i.e. during the period of higher efficacy of metastatic ability of the tumour, it is maintained at a good level when NAMI-A is given during late phases of tumour growth, i.e. when the metastatic release from the primary tumour is virtually complete, and it is practically zero when NAMI-A is given during the very early phases of tumour growth when metastasis formation is in a very early phase (Table 4).

Although a direct tumour cytotoxicity is not expected to be responsible for the anti-metastatic effects observed in mice, NAMI-A modifies some parameters

Table 4. Effects of NAMI-A on MCa mammary carcinoma at different stages of growth

Treatment	Body weight[b]	Primary tumour[c]	Lung metastases[a]			
			Number/animal		Weight (mg)/animal	
		mean±S.E.	mean±S.E.	% Var.[d]	mean±S.E.	% Var.[d]
Controls	–	1.00±0.02	0	–	0	–
NAMI-A 1–5	–3.8%	0.85±0.03[e]	0	–	0	–
Controls	–	1.38±0.04	6.75±1.25	–	27.6±16.1	–
NAMI-A 6–10	–2.8%	1.10±0.06[e]	5.25±2.98	–22	25.0±16.2	–9
Controls	–	2.61±0.11	41.8±5.2	–	976±197	–
NAMI-A 11–15	–3.5%	2.11±0.07[e]	11.4±3.7	–73[f]	66±23	–93[f]
Controls	–	4.03±0.31	ND	–	ND	–
NAMI-A 16–20	–2.8%	3.56±0.51[e]	26.0±8.9	–38[e]	416±222	–57[e]

Groups of 7 mice implanted i.m. in the calf of the left hind leg with 10^6 tumour cells on day 0, were given i.p. 35 mg/kg/day NAMI-A in the inclusive days reported in table. The primary tumour was surgically removed 24 h after last dosage, and lung metastases were measured on day 22. Data on metastases for the group NAMI-A 16–20 was calculated vs controls of the group NAMI-A 11–15.
a Measured after sacrifice of the animals.
b Percentage variation between the beginning and the end of the treatment with respect to the untreated control group.
c Measured at surgical removal of the primary tumour.
d Percentage variation with respect to the untreated control groups.
e $p < 0.05$ statistically different from the controls.
f $p < 0.01$ statistically different from the controls.

of the cell cycle of tumour cells. Following staining of primary tumour cells with propidium iodide, NAMI-A, unlike NAMI, causes a significant increase in the number of cells in the G_2/M phase vs controls in a study performed in CBA mice carrying MCa mammary carcinoma and treated i.p. with 35 mg/kg/day NAMI-A for six consecutive days starting on day 9 from tumour implantation. NAMI-A also significantly reduces the DNA index from 1.589±0.010 to 1.513±0.020 [49]. Besides this effect, another one can be detected: ex vivo samples of MCa mammary carcinoma contain two different cell populations of tumour cells, one characterised by a diploid cycle and identified by the positivity to pan leukocyte marker CD45, and the other (CD45 negative) characterised by the aneuploid cycle. NAMI-A reduces the amount of diploid cells in the primary tumour mass without affecting their distribution in the cell cycle phases [49]. The effects on the immune system, after in vivo treatment with NAMI-A, were also investigated in the spleen of BD2F1 mice bearing Lewis lung carcinoma. NAMI-A significantly increases the percentage of CD8$^+$ cells at all dose levels (three dose levels), while CD4$^+$ cells increase at the lowest dose and remain unchanged at medium and high doses. Globally, the ratio between CD4$^+$ and CD8$^+$ cells is similar in controls and treated groups, except at the highest dose at which it is reduced from 1.4±0.1 to 0.8±0.1.

On histological examination, the effects on primary tumour growth are of the same type as those reported for NAMI, but markedly more pronounced, and consist of a significant increase in the thickness of the connective tissue of tumour capsule and around the tumour blood vessels.

NAMI-A is also markedly more effective than cisplatin and dacarbazine in the control of post-surgical lung metastases in mice bearing MCa mammary carcinoma (Table 5). Compared to cyclophosphamide, NAMI-A is less effective but, unlike cyclophosphamide which also reduces by 12% body weight gain during treatment and by 55% spleen weight, NAMI-A is completely devoid of toxicity to the host. The results of these experiments clearly provide evidence for the different behaviour of NAMI-A, compared to cisplatin, and for the control of lung metastases of Lewis lung carcinoma, stressing the proposed role of NAMI-A as a selective anti-metastasis agent. In other words, the pronounced reduction of lung metastasis weight, greater than that of metastasis number, in addition to the results of Table 4, show that NAMI-A is endowed with the desirable property of reducing metastases rather than inhibiting their formation.

The fate of NAMI-A in healthy CBA mice after a single i.v. treatment of 200 mg/kg was determined by measuring, by atomic absorption spectroscopy, the ruthenium content in the blood, the kidneys, the liver, the lungs, the small and large bowel, the spleen, the inguinal lymph nodes and the brain (Fig. 6). The ruthenium concentration in the blood rapidly falls and, 5 min after i.v. treat-

Table 5. Effects of NAMI-A, DTIC, cyclophosphamide and cisplatin on Lewis lung carcinoma

Treatment	Body weight[b]	Spleen weight[c]	Lung metastases[a]			
			Number/animal		Weight (mg)/animal	
			mean±S.E.	% Var.[d]	mean±S.E.	% Var.[d]
Controls	–	–	14.8±0.6	–	985±41.0	–
NAMI-A	1%	–11%	4.2±0.5	–72	73.3±20.0	–93
DTIC	–3%	–18%	12.2±1.6	–18	765±139	–22
Cyclophos-phamide	–12%	–55%	0	–100	0	–100
Cisplatin	–11%	–52%	13.5±0.6	–9	865±73.0	–12

Groups of 7 mice implanted i.m. in the calf of the left hind leg with 10^6 tumour cells on day 0 were given i.p. 35 mg/kg/day NAMI-A, 40 mg/kg/day cyclophosphamide, 60 mg/kg/day DTIC and 2 mg/kg/day cisplatin on days 12–17, after surgical removal of the primary tumour on day 11. Lung metastases were measured on day 25
a Measured after sacrifice of the animals.
b Percentage variation between the beginning and the end of the treatment with respect to the untreated control group.
c Percentage variation of the spleen weight with respect to the one of the untreated control group, measured when the animals were killed.
d Percentage variation with respect to the group of untreated controls.

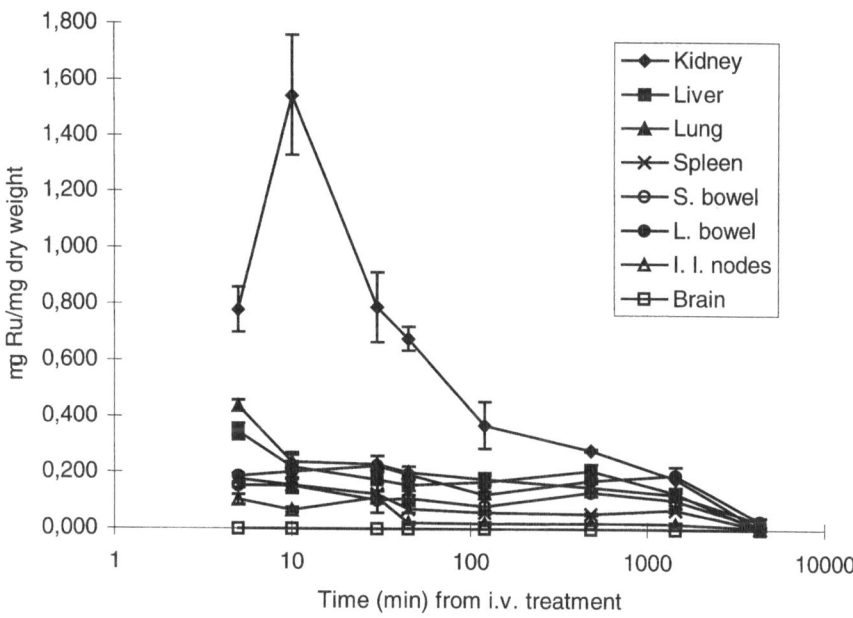

Fig. 6. Concentration of ruthenium in the organs and tissues examined. Mice treated i.v. with NAMI-A at t=0 were killed by a lethal dose of Ketalar at 5, 10, 30, 45, 120, 480, 1440 and 4320 min. Ruthenium concentration is expressed as μg/ml (mean±S.E.) of individual samples obtained from 4 independent mice

ment, is about 100 μg/ml, corresponding to less than 10% of total ruthenium administered. During the same period, the ruthenium concentration in urine is about 20 mg/ml. In the kidney, the ruthenium concentration is markedly higher than in any other tissue analysed, peaking 10 min after treatment. No ruthenium was detected in the brain.

The pharmacokinetic parameters, calculated either considering a mono- or a bi-compartment model, are equivalent: $t_{1/2}$ is approximately 12 h with a clearance of 1.4 ml/h; the volume of distribution is approximately of 26 ml with an AUC of 550 mg/l/h. Ruthenium appears to be retained in three main areas, the blood, the kidneys and the liver. The amount of ruthenium retained by the body after 2 h of treatment is about 85% of the administered dose. These data show that NAMI-A is rapidly cleared from the blood compartment immediately after i.v. administration and in the urine its concentration is three orders of magnitude greater than that of plasma. Apparently there is no differential accumulation of ruthenium in the lungs that might account for a selective anti-metastatic effect due to a cytotoxic concentration in this site, nor in any other specific organ examined.

Because of its activity, which is very similar to that of NAMI, it is possible to credit the novel compound NAMI-A with all the characteristics already described in detail for NAMI. Therefore, NAMI-A turns out to be a novel anti-tumour compound. It is distinguished from any other anti-tumour drug presently used for its selective action on lung metastases of solid tumours, since its mechanism is not related to a direct cytotoxicity for tumour or metastatic cells. The reduction of metastatic tumour is followed by a significant prolongation of the life-span of the tumour bearing hosts, even longer if drug treatment is accompanied by surgical removal of the primary tumour or by the administration of other cytotoxic compounds such as cisplatin or 5-fluorouracil (G Sava, data on file).

3
Conclusions

The first, and perhaps most important, conclusion that can be drawn from our extensive work in this field is that a number of ruthenium(II) and ruthenium(III)–Me_2SO compounds are endowed with anti-tumour activity against experimental models of murine tumours. This activity is, in several cases, different from that of platinum drugs; the main difference concerns the selectivity of effects against metastases of solid metastasising tumours. Although surgical techniques and adjuvant therapies have reached a remarkable level of development, a large proportion of patients inevitably suffer from the formation of metastases from primary lesions. Eradication of metastases in patients with malignant tumours is thus a very important goal in clinical oncology.

Besides sharing the common feature of bearing dimethylsulfoxide ligands, the complexes examined over the years did not indicate a clear structure-activity relationship that might be usefully exploited in further synthetic efforts. Apparently, the presence of simple N-donor ligands on ruthenium(III) species is a requirement for finding favourable biological responses. One of these complexes, NAMI-A, has undergone advanced preclinical trials and might be further developed as an anti-metastatic drug. The ease of reduction of NAMI-A in aqueous solution suggests that this process is likely to occur, at least partially, also in vivo. Therefore, both ruthenium(III) and ruthenium(II) species might be present in vivo after administration of NAMI-A, each species undergoing independent hydrolytic processes. Experiments are in progress to assess which species is/are responsible for the biological activity of the precursor.

The mechanism of action of the Ru–Me_2SO complexes (assuming that they share a common mechanism) is still largely unknown, but experimental evidence indicates that a direct cytotoxic action, similar to that of platinum drugs, can be excluded. It has been demonstrated that active ruthenium complexes interact with DNA in vitro and induce local conformational alterations different from those induced by cisplatin. However, in good agreement with the documented cytotoxicity and systemic toxicity of ruthenium compounds, which is lower than that of platinum compounds, DNA binding was observed only at relatively high concentrations, significantly larger than those at which cisplatin produces similar effects. Therefore, even though DNA may represent a target for the action of ruthenium complexes, this hypothesis might become realistic in vivo only if substantial amounts of ruthenium are able to enter cells and reach nuclear DNA. In this regard, experimental evidence obtained by us and by others suggests that the delivery (and accumulation) of ruthenium drugs in cells might be mediated by transferrin-transport and by activation-by-reduction mechanisms. However, it must be stressed that cytotoxicity, when present, is not simply related to the amount of ruthenium that enters tumour cells. Concerning the selectivity of the effects of NAMI-A for the lung metastases of the solid tumours used, it can be excluded that it depends on a higher concentration of this compound in the lungs, where, following a cycle of administration at active doses, its concentration is always around 10^{-4} M, a concentration totally inactive when applied in vitro on tumour cells of different origin (G Sava, unpublished data).

Finally, we wish to add a word of caution in regard to the screening procedures for inorganic anti-tumour compounds. Such procedures have been developed following the experience accumulated with platinum drugs and are mainly focused on the cytotoxic activity of the tested compounds. If on the one hand it is difficult to discuss experimental models predictive of clinical activity, on the other it should be stressed that some experimental situations are much closer than others to those usually encountered in humans. Compounds that fail to induce cytotoxicity in vitro are normally discarded and do not undergo further investigations in vivo. However, our experience shows that non-cytotoxic metal complexes can have a remarkable anti-neoplastic activity and, most importantly, an activity different from that of clinically used platinum drugs. In other words, we believe that screening procedures based on the assessment of cytotoxicity are likely to select compounds that are similar to already used platinum drugs, while they are very likely to discard what they are supposedly looking for, i.e. new compounds with a mechanism of action and a spectrum of activity different from those of platinum drugs.

References

1. (a) Barnard CFJ (1989) Platinum Metal Rev 33:162; (b) Reedijk J (1996) Chem Commun 801 and references cited therein
2. Takahara PM, Rosenzweig AC, Frederick CA, Lippard SJ (1995) Nature 377:649
3. (a) Cleare MJ, Hydes PC (1980) Antitumor properties of metal complexes. In: Siegel H (ed) Metal ions in biological systems. Marcel Dekker, New York, p 1; (b) Farrell N (1989)Transition metal complexes as drugs and chemotherapeutic agents. Kluwer Academic Publishers, Dordrecht, The Netherlands; (c) Haiduc I, Silvestru C (1990) Coord Chem Rev 99:253; (d) Keppler BK (ed) (1993) Metal complexes in cancer chemotherapy. VCH, Weinheim
4. (a) Gielen M (1994) Metal Based Drugs 1:213; (b) Gielen M (1995) Metal Based Drugs 2:99
5. Sadler PJ, Sue RE (1994) Metal Based Drugs 1:107
6. (a) Köpf-Maier P (1993) Antitumor bis(cyclopentadienyl)metal complexes. In: Keppler BK (ed) Metal complexes in cancer chemotherapy. VCH, Weinheim, p 259; (b) Keppler BK, Friesen C, Vongerichten H, Vogel E (1993) Budotitane, a new tumor-inhibiting ti-

tanium compound: preclinical and clinical development. In Keppler BK (ed) Metal complexes in cancer chemotherapy. VCH, Weinheim, p 297

7. Clarke MJ (1993) Ruthenium complexes: potential role in anti-cancer pharmaceuticals. In: Keppler BK (ed) Metal complexes in cancer chemotherapy. VCH. Weinheim, p 129
8. Clarke MJ (1980) Oncological implications in the chemistry of ruthenium. In: Sigel H (ed) Metal ions in biological systems. Marcel Dekker, New York, p 231
9. Sava G, Zorzet S, Giraldi T, Mestroni G, Zassinovich G (1984) Eur J Cancer Clin Oncol 20:841
10. Keppler BK, Rupp W (1986) J Cancer Res Clin Oncol 111:166
11. Alessio E, Mestroni G, Nardin G, Attia WM, Calligaris G, Sava G, Zorzet S (1988) Inorg Chem 27:4099
12. Sava G, Pacor S, Mestroni G, Alessio E (1992) Clin Exp Metastasis 10:273
13. (a) Fricker SP (1995) Platinum Metal Rev 39:150; (b) Davies NA, Wilson MT, Slade E, Fricker SP, Murrer BA, Powell NA, Henderson GR (1997) Chem Commun 47
14. (a) van Vliet PM, Toekimin SMS, Haasnoot JG, Reedijk J, Novakova O, Vrana O, Brabec V (1995) Inorg Chim Acta 231:57; (b) Novakova O, Kasparkova J, Vrana O, van Vliet P M, Reedijk J, Brabec V (1995) Biochemistry 34:12369
15. Vilaplana R, Romero MA, Quiros M, Salas JM, Gonzalez-Vilchez F (1995) Metal Based Drugs 2:211
16. Clarke MJ, Bailey VM, Doan PE, Hiller CD, LaChance-Galang KJ, Daghlian H, Mandal S, Bastos CM, Lang D (1996) Inorg Chem 35:4896
17. Mestroni G, Alessio E, Sava G, Pacor S, Coluccia M, Boccarelli A (1994) Metal-Based Drugs 1:43
18. Mestroni G, Alessio E, Calligaris M, Attia WM, Quadrifoglio F, Cauci S, Sava G, Zorzet S, Pacor S, Monti-Bragadin C, Tamaro M, Dolzani L (1989) Chemical, biological and antitumor properties of ruthenium(II) complexes with dimethyl sulfoxide. In: Clarke MJ (ed) Progress in clinical biochemistry and medicine. Springer, Berlin Heidelberg New York, p 71
19. Sava G, Pacor S, Zorzet S, Alessio E, Mestroni G (1989) Pharmacol Res 21:617
20. Alessio E, Balducci G, Lutman A, Mestroni G, Calligaris M, Attia WM (1993) Inorg Chim Acta 203:205
21. Keppler BK, Henn M, Juhl UM, Berger MR, Niebl R, Wagner FE (1989) In: Progress in clinical biochemistry and medicine Springer, Berlin Heidelberg New York, p 41
22. Pacor S, Sava G, Ceschia V, Bregant F, Mestroni G, Alessio E (1991) Chem-Biol Interactions 78:223
23. Sava G, Pacor S, Mestroni G, Alessio E (1992) Anticancer Drugs 3:25
24. Mestroni G, Alessio E, Sava G, Pacor S, Coluccia M (1993) In: Keppler BK (ed) Metal complexes in cancer chemotherapy. VCH, Weinheim, p 157
25. Sava G, Pacor S, Bergamo A, Cocchietto M, Mestroni G, Alessio E (1995) Chem-Biol Interactions 95:109
26. Sava G, Pacor S, Coluccia M, Mariggiò M, Cocchietto M, Alessio E, Mestroni G (1994) Drug Invest 8:150
27. Gagliardi R, Sava G, Pacor S, Mestroni G, Alessio E (1994) Clin Expl Metastasis 12:93
28. Bergamo A, Cocchietto M, Capozzi I, Mestroni G, Alessio E, Sava G (1996) Anticancer Drugs 7:697
29. Capozzi I, Clerici K, Cocchietto M, Salerno G, Bergamo A, Sava G (1998) Chem-Biol Interact 113:51
30. Kreuser ED, Keppler BK, Berdel WE, Piest A, Thiel E (1992) Semin Oncol 19:73
31. Puccetti P, Romani L, Fioretti MC (1985) Trends in Pharm Sci 6:213
32. Sava G, Zorzet S, Pacor S, Mestroni G, Zassinovich G (1989) Cancer Chemother Pharmacol 24:302
33. Sava G, Salerno G, Bergamo A, Cocchietto M, Gagliardi R, Alessio E, Mestroni G (1996) Metal Based Drugs 3:67

34. Sava G, Capozzi I, Bergamo A, Gagliardi R, Cocchietto M, Onisto M, Alessio E, Mestroni G, Garbisa S (1996) Int J Cancer 68:60
35. New salts of anionic complexes of Ru(III) as antimetastatic and antineoplastic agents. PCT. C 07F 15/00, A61K 31/28. WO 98/00431 08.01.98
36. Geremia S, Alessio E, Todone F (1996) Inorg Chim Acta 253:87
37. Alessio E, Balducci G, Calligaris M, Costa G, Attia WM, Mestroni G (1991) Inorg Chem 30:609
38. Calligaris M, Carugo O (1996) Coord Chem Rev 153:83
39. Keppler BK, Rupp W, Juhl UM, Enders H, Niebl R, Balzer W (1987) Inorg Chem 26:4366
40. Ni Dhubhghaill OM, Hagen WR, Keppler BK, Lipponer K-G, Sadler PJ (1994) J Chem Soc Dalton Trans. 3305
41. Anderson C, Beauchamp AL (1995) Inorg Chem 34:6065
42. Brabec V, Malina J, Alessio E unpublished results
43. Gallori E, Vettori C, Alessio E, Gonzalez-Vilchez F, Orioli P, Dal Poggetto G, Messori L unpublished results
44. Messori L, Kratz F (1994) Metal Based Drugs 1:161
45. Messori L, Kratz F, Alessio E (1996) Metal Based Drugs 3:1
46. Kratz F, Keppler BK, Messori L, Smith C, Baker EN (1994) Metal Based Drugs 1:169
47. Smith CA, Sutherland-Smith AJ, Keppler BK, Kratz F, Baker EN (1996) J Bioinorg Chem 1:424
48. Frasca D, Ciampa J, Emerson J, Umans RS, Clarke MJ (1996) Metal Based Drugs 3:197
49. Sava G, Capozzi I, Clerici K, Gagliardi R, Alessio E, Mestroni G (1998) Clin Exp Metastasis 16:371

Non-Platinum Antitumor Compounds

Thomas Pieper, Karl Borsky, Bernhard K. Keppler

Institute of Inorganic Chemistry, University of Vienna, Waehringerstrasse 42, A-1090 Vienna, Austria
E-mail: keppler@merlin.ap.univie.ac.at

Non-platinum antitumor compounds have gained increasing interest during the last decade. A variety of antitumor active agents, from early transition to main-group metal or metalloid compounds, has been found and evaluated. Due to their different chemical characteristics, the mode of action and spectrum of activity of these compounds differ from each other. Antitumor activity in vitro is known for many compounds. Some compounds, like the ruthenium complexes, also exhibit promising in vivo activity. Two early transition metal complexes show interesting activity especially in experimental colon tumor models and are under evaluation in clinical trials: titanocene dichloride and budotitane. The two germanium complexes spirogermanium and germanium-132 have been evaluated but, despite their moderate toxicity, no encouraging results have been obtained. Gallium nitrate and gallium chloride are active against lymphomas and bladder cancer, and show a positive effect on hypocalcaemias caused by tumors. They have undergone a number of clinical trials and are being investigated in promising combination regimens. Arsenic trioxide was found to be active against a rare blood cancer in clinical studies in China.

Keywords. Antitumor compounds, Metal complexes, Tumor inhibitors, Cytotoxic compounds, Cancer therapy

List of Abbreviations

2,5-diMeBo	2,5-dimethylbenzoxazole
DNA	deoxyribonucleic acid
2-dqmp	diethyl-2-quinolylmethylphosphonate
en	1,2-diaminoethane
ESR	electron spin resonance
Hpm	2-hydroxymethylpyridine
ID	initial dose
im	imidazole
ind	indazole
2-MeBo	2-methylbenzoxazole
mbpa	methyl-4,6-O-benzylidene-n-deoxy-n-(diphenylphosphino)-α-D-altropyranoside
N-MeIm	N-methylimidazole
phen	phenanthroline
pm	2-oxymethylpyridine
terpy	2,2':6'2"-terpyridine

1
Introduction

Since the accidental discovery of the tumor-inhibiting properties of cisplatin, every effort has been made to develop new platinum complexes and elucidate the mode of action of these inorganic antitumor agents. A lot of excellent scientific work has been carried out and, of course, a lot of money and human resources have been expended to find another breakthrough in tumor therapy with platinum compounds.

Today, cisplatin is a well-established chemotherapeutic drug, especially against testicular carcinomas, but expansion to a different or broader antitumor spectrum has not been obtained with cisplatin or its direct analogues such as carboplatin.

The main improvement achieved with these compounds is reduced systemic toxicity or a different toxicity profile and, in some cases, moderate activity against cisplatin-resistant tumors. New developments like orally administrable (JM216) platinum complexes or combination therapy (oxaliplatin/5-fluoruracil, approved for the treatment of colon cancer in France) are promising, but whether these treatment regimens can be established in clinical practice remains to be seen.

Therefore, attention is being more and more directed towards non-platinum antitumor metal compounds. A wide-spread variety of antitumor active agents, from early transition metal to main group element compounds, has been found and evaluated. Because of their different chemical characteristics, the mode of action and spectrum of activity of these compounds should differ significantly. They possess totally different redox behavior or hydrolysis reactions and kinetics. Preferences for distinct biologically occurring nucleophiles determine a different interaction with biological targets. Of what nature is binding to nucleobases? Is interaction with DNA the crucial point, as it is thought to be with platinum complexes? Which protein binding behavior is exhibited?

The chemistry of titanocene derivatives is different from the chemistry of ruthenium or gold complexes. A relationship between structurally cisplatin-related gold or palladium complexes, with *cis*-standing chloride ligands, does not automatically lead to the same mode of action as, for example, some gold(III) compounds are reduced to gold(I) species and undergo fast hydrolysis under physiological conditions.

As a matter of fact, less is known about the mode of action of most non-platinum compounds compared to platinum antitumor drugs. The efforts that have been made to elucidate the biological action of ruthenium or titanium antitumor active complexes, for example, are not comparable with those for platinum complexes. Additionally, too much might be considered to be analogous to the field of platinum.

In the following, the antitumor properties of metal or metalloid compounds of elements spread all over the periodic table will be summarized and described. Less will be reported about investigations towards their mode of action.

Certainly, this aspect should be kept in mind when synthesizing new complexes. Even more important are some "practical" aspects. Unfortunately, many interesting antitumor active compounds cannot be introduced into clinical trials because of insufficient water solubility. Adding solubilizers and carrying out sophisticated galenic procedures with complexes insoluble in water usually causes additional problems in the analytics of the formulation and hence registration with the national drug authorities. Often it is not even possible to achieve sufficient concentrations of the insoluble complexes in aqueous solution with the help of solubilizers. Furthermore, hydrolytic instability and a variety of decomposition species hinder a clinical application. Metal complexes are often more sensitive towards solvents, humidity, light or air than organic pharmaceuticals, and it seems to be difficult to overcome prejudiced attitudes of pharmaceutical companies, even if the handling of a compound is feasible. Moreover, application of a regimen containing arsenic, for example, might not be that "popular".

Another obstacle for clinical development of non-platinum antitumor complexes might be that the compounds have often not been investigated in realistic and sophisticated tumor models.

In view of the fact that the P388 leukemia tumor model is very sensitive to platinum compounds, for example, it would be surprising to find other complexes showing much better effects than cisplatin.

It must be taken into account that transplantable tumors like the P388 model, just as other primary screening methods, are more appropriate for separating active complexes from inactive ones rather than for assessing the status of different groups of substances such as platinum and ruthenium complexes. Hence, it is by no means an exclusive criterion if tumor-inhibiting complexes do not show a higher activity in this model than cisplatin does. They may well qualify for further development. Within a class of compounds, however, a transplantable tumor is fairly well suited for distinguishing activity grades, because all the derivatives from a class of substances probably act on the basis of a similar molecular mechanism. When a highly active structure type is found among the compounds, it is necessary to resort to sophisticated and clinically meaningful models, like xenografts or, in particular, autochthonously growing cancer models, in order to be able to assess the potential advantages of therapy with a particular derivative. Here autotopically transplanted human tumor xenografts and autochthonous tumors are in question.

In general, one has to be aware of the fact that metal complexes are often cytotoxic in vitro at a significantly lower dose than organic drugs. Thus, it is necessary to assess in vitro findings cautiously. The potential of a metal complex as an antitumor therapeutic agent finally has to be evaluated definitely by adequate in vivo investigations.

2
Main-Group Metal Compounds

2.1
Group 13 Metal Compounds – Ga, In, Tl

Gallium salts were examined in preclinical studies along with analogous compounds of aluminum, indium, and thallium [1]. All compounds showed in vivo activity in the Walker 256 carcinosarcoma when treatment was carried out intraperitoneally. However, all of them showed no or very little activity in the leukaemias L1210, K1964 and P388 and in the Ehrlich ascitic carcinoma. In systemic therapy, i.e. when the substance is not injected directly into the tumor, only gallium nitrate and indium nitrate showed activity, particularly in the intraperitoneal therapy of subcutaneously transplanted Walker 256 carcinosarcoma. Indium nitrate, however, is much more toxic than the corresponding gallium compound.

2.1.1
Gallium

The antitumor activity of gallium salts is quite independent of the anion. Besides gallium nitrate, gallium chloride and citrate are also frequently used. The fact that gallium nitrate shows better antitumor effects than the other group 13 element compounds might be due to its higher concentration in experimental tumors. Gallium accumulates in particular tumors, for example, the Walker carcinosarcoma 256, lymphosarcoma P-1798 and adenocarcinoma 755 [2]. These findings were achieved in experiments with the radionuclide Ga-57. This radionuclide is used in the clinic for scintigraphic purposes, particularly for the detection of lymphomas, bone tumors and bone metastases. It is supplied in a solution containing high levels of citrate to suppress hydrolysis to $Ga(OH)_3$.

As gallium(III) resembles iron(III) in certain respects (e.g. ionic radii: 62 pm and 65 pm), it is able to bind to the iron transport protein transferrin and is incorporated into cells, especially tumor cells (because of the large number of transferrin receptors on tumor cell surface), by transferrin receptor dependent and independent pathways [3–5]. Additionally, gallium(III) can be incorporated into the iron storage protein ferritin [6]. The development of gallium resistance might be due to a down regulation in the transport of gallium into cells. In parallel, iron transport is affected, causing changes in the regulation of ferritin gene expression [7]. The mechanism of DNA synthesis and cell growth inhibition by gallium nitrate might be a combination of a blocked iron availability to ribonucleotide reductase and an inhibition of the enzyme itself [8].

In phase I clinical studies nephrotoxicity was found to be the dose-limiting side effect of gallium nitrate. In addition, there were gastrointestinal side effects and temporary hypercalcaemia. In phase II studies, activity of gallium nitrate in Hodgkin's and non-Hodgkin's lymphomas and bladder cancer was demonstrated [9–11].

Gallium salts have a positive effect on hypocalcaemias caused by tumors, restoring calcium levels to normal [12–15]. In comparative clinical trials it proved a more effective antihypercalcaemic agent than calcitonin or etidronate [16,17].

A combination of gallium chloride, cisplatin and VP16 showed improved results in patients with non-small-cell lung tumors compared to a therapy with only cisplatin and VP16 [18].

Recently, phase II clinical studies have focused on combination regimens consisting of gallium nitrate, vinblastine and ifosfamide (and filgrastim) [19–21]. The results in patients with advanced carcinoma of the urothelium or ovarian cancer have encouraged further evaluation of this regimen, whereas a combination of gallium nitrate with 5-fluoruracil has proved unfavorable [22].

Remarkable reports have been published about synergistic effects. As gallium effects ribonucleotide reductase, it can interact synergistically with other inhibitors of this enzyme, for example, hydroxyurea. Gallium nitrate, administered intravenously, and hydroxyurea, administered orally, have significant activity in non-Hodgkin's lymphoma [23]. The lack of cross-resistance between gallium nitrate and hydroxyurea and other antineoplastic agents can lead to new combinations with these drugs. Synergistic effects were found in preclinical trials with fludarabine and α-interferon, on the proliferation of human leukemic HL60 cells and on T-lymphoblastic leukemic CCRF-CEM cells, respectively [24,25]. As ribonucleotide reductase activity increases in the S phase of the cell cycle, gallium should be most effective in this phase. Combination with a drug that effects a different phase of the cell cycle and attacks another target might be effective. Gallium nitrate and taxol (paclitaxel, arrests cells in mitosis) is such a combination. A drug-schedule-dependent interaction was found against human breast carcinoma cells [26]. The best synergism was found when gallium nitrate was added first, followed by taxol 16 to 24 h later. Taxol before gallium nitrate or simultaneous treatment was less effective.

In order to improve the bioavailability and increase the plasma concentration of gallium in oral administration, gallium complexes containing organic ligands have been synthesized [27]. The necessary plasma concentration can easily be reached with tris(8-quinolinolato)gallium(III) (Fig. 1). It shows improved bioavailability compared to gallium chloride but also a greater toxicity [28]. Tris(8-quinolinolato)gallium(III) exhibits better antiproliferative and antihypercalcaemic properties than gallium nitrate in the Walker carcinosarcoma 256 in a dose of 24 mg/kg, where no severe side effects on liver or kidney were observed [29]. It has a threefold capacity of apoptosis induction in vitro compared to gallium nitrate [30].

2.2
Group 14 Metal Compounds – Ge, Sn

Two germanium compounds, spirogermanium and germanium-132, have undergone clinical trials but, despite moderate toxicity, the results are not very encouraging.

In general tin compounds, especially compounds containing a diorganotin moiety, showed promising activity in vitro that could not be confirmed in solid tumors in vivo.

2.2.1
Germanium

The organometallic compound spirogermanium, 8,8-diethyl-2-[3-(N-dimethylamino)-propyl]-2-aza-8-germaspiro[4,5]decane (Fig. 2), containing an azaspir-

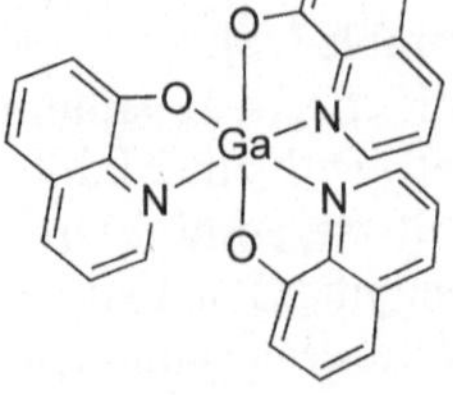

Fig. 1. Tris(8-quinolinolato)gallium(III)

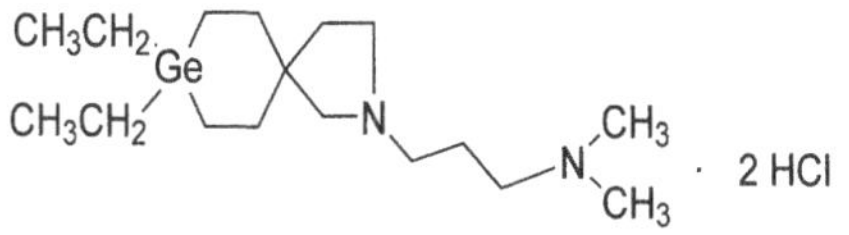

Fig. 2. Spirogermanium

ane ring structure, showed cytotoxic activity in numerous cell cultures. It also inhibited the growth of Walker 256 sarcoma, mammary adenocarcinoma 13762 and prostatic carcinoma 11095 in rats [31]. Only a minor effect was found on the leukaemias L1210 and P388 and the melanoma B16. Spirogermanium is not cell-cycle-phase specific. It inhibits the synthesis of DNA, RNA and proteins in vitro and inhibition of protein synthesis seems to be its main mode of action [32,33].

Numerous phase II clinical studies have been carried out with different tumor types such as ovarian carcinomas, kidney tumors, glioblastomas, colon tumors, carcinomas of the prostate, malignant lymphomas, non- and small-cell lung carcinomas, carcinomas of the cervix, mammary carcinomas, and melanomas [34–38].

Despite the promising moderate toxicity, mainly neurotoxicity was observed in phase I clinical studies; no sufficient tumor-inhibiting activity could be found in phase II studies that would recommend further clinical use.

The second germanium compound evaluated in clinical trials is the polymeric germanium sesquioxide germanium-132, (bis[(carboxyethyl)germanium]trioxide, $[(GeCH_2CH_2COOH)_2O_3]_n$, which showed preclinical activity in ascitic hepatomas AH44 and AH46, experimental bladder tumors (BC47), and in the Lewis lung carcinoma [39,40]. Based on animal experiments it is assumed that germanium-132 has a stimulating effect on the immune system, along with interferon-inducing properties [41–43].

Germanium-132 has been under clinical trials, especially in Japan, but no really promising activity has been found [44–50].

Antitumor activity has been reported for some organogermanium sesquioxide and sesquisulphide analogues [51,52], as well as of some decaphenylgermanocene compounds [53].

2.2.2
Tin

A series of diorganotin dihalides with nitrogen donor ligands has been screened for activity against P388 lymphotic leukemia. The activity of these organotin compounds is controlled by the diorganotin moiety, whereas hydrolysis of the halide ligands is important for the transport to the site of action [54]. Tin(IV) compounds, like the octahedral diorganotin dihalide complexes $[R_2SnX_2L_2]$ (R= alkyl, phenyl; X=F, Cl, Br, I, NCS; L=unidentate ligand like pyridine), were found to be active against P388 leukaemia but not in other, more meaningful, tumor models and are not under current investigation [55–57].

Di-*n*-butyltin 2,6-pyridine dicarboxylates were found to be more active than cisplatin against mammary tumor (MCF-7) and colon carcinoma (WiDr) cell

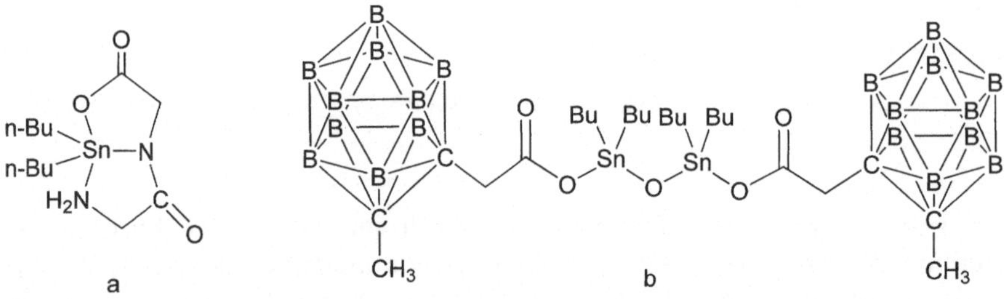

Fig. 3. *a* Di-*n*-butyltin(IV)glycylglycinate, *b* bis[(2-methyl-1,2-dicarba-*closo*-dodecaborane-1-carboxylato)-di-*n*-butyltin]oxide

lines [58,59]. Some di-*n*-butyltin difluorbenzoates also showed interesting activity [60], whereas tri-*n*-butyltin difluorbenzoates are less active than the di-*n*-butyltin and the triphenyltin derivatives [61].

Triphenyltin carboxylates were active against MCF-7(breast) and WiDr (colon) cell lines and reached interesting, low initial dose (ID) values compared to clinically established antitumor drugs [62].

Another structure type is represented by di-*n*-butyltin(IV)glycylglycinate (Fig. 3a) which reaches T/C values of about 150% in the P388 leukaemia (T/C: median survival time of treated animals vs. median survival time of untreated control animals · 100). Only marginal activity was found in other tumor systems [63–65].

Antitumoral activity has been found for a structurally interesting new class of di-*n*-butyltin oxides. Dimeric bis[(2-methyl-1,2-dicarba-*closo*-dodecaborane-1-carboxylato)-di-*n*-butyltin]oxide (Fig. 3b), with a central $Bu_4Sn_2O_2$ unit and carboranecarboxylate ligands, is more active in vitro against the human tumor cell lines MCF-7 (breast), EVSA-T (breast), WiDr (colon), IGROV (ovarian), M19 MEL (melanoma), A498 (renal) and H226 (non-small-cell lung cancer) than 5-fluoruracil, cisplatin and carboplatin but less active than methotrexate and doxorubicin [66].

2.3
Group 15 Metal Compounds – As, Sb, Bi

Interesting results in recent clinical trials were obtained with arsenic trioxide in a rare blood cancer. In contrast to their significance in other therapeutic fields, only a few compounds of antimony and bismuth are known to have antitumor activity.

2.3.1
Arsenic

The development of the organoarsenic drug Salvarsan by Paul Ehrlich established chemotherapy. However, since the early days of chemotherapy at the be-

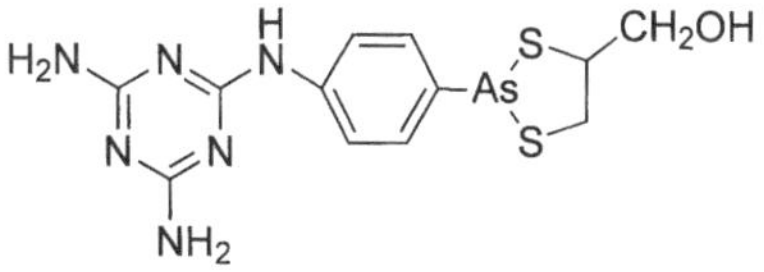

Fig. 4. Melarsoprol

ginning of the century, no arsenic compound has achieved any comparable significance. A promising new development has started recently, which has introduced a classical inorganic compound, arsenic trioxide, into cancer therapy. Interestingly, the "new" drug was discovered in a "screening" of traditional Chinese remedies during the cultural revolution [67]. Arsenic trioxide was discovered to be a potent agent against acute promelocytic leukemia (APL, a rare blood cancer). It was found to be more effective than the established therapeutic all-trans retinoic acid. Clinical studies in China demonstrated that As_2O_3 can induce remission in patients with APL by inducing apoptosis, without differentiation in retinoic acid sensitive or resistant leukemia cells [68–71]. The apoptosis induction occurs independently to the retinoid pathway. Very recently indication of a synergism between arsenic trioxide and retinoic acid has been found [72].

The organoarsenic drug melarsoprol (Fig. 4, a drug used for treatment of trypanosomiasis) inhibits the growth of lymphoid leukemic cells by promoting apoptosis at lower concentrations than As_2O_3 [73,74]. Clinical investigations against acute and chronic myeloid and lymphoid leukemias have been initiated [75].

2.3.2
Antimony, Bismuth

The use of antimony and bismuth compounds as antimicrobial agents is well known (see Sadler, TBIC, Vol 2). In contrast, only a few compounds of these group 15 elements are known to have antitumor activity.

Some organometallic bismuth(III) thiolates exhibit antitumor activity [76]. Bismuth(III) complexes of different thiosemicarbazone derivatives inhibit growth of tumor cells in SW948 (rectum) and SW707 (colon) cell lines [77] and show antiproliferative activity against clonogenic cells from freshly explanted human tumors [78].

Administration of bismuth nitrate prior to chemotherapy with cisplatin may be effective in the treatment of advanced bladder tumors [79,80].

Diphenylantimony(III) dithiophosphorus complexes are active against Ehrlich ascites tumor cells in mice [81–83]; the most active compound is (diisopropylphosphorodithioato)diphenylantimony(III) [84].

3
Transition Metal Compounds

3.1
Early Transition Metal Complexes – Ti, Zr, Hf, V, Nb, Mo, Re

Among the early transition metal compounds the metallocene complexes are of special interest. The titanium derivative is under clinical evaluation. Another titanium complex, budotitane, that exhibited excellent results in experimental colon cancer models, has also undergone clinical trials. In general, the fast kinetics and decomposition in aqueous solution of titanium complexes make a formulation for the clinic difficult.

Among others, rhenium-186 and technetium-99m compounds play an important role as radiopharmaceuticals and are well known as diagnostic radio imaging agents. Rhenium-186 hydroxyethylidene bisphosphonate, for example, is used for the palliative treatment of metastatic bone pain [85,86], rhenium-186-labeled monoclonal antibodies against ovarian carcinoma or gastrointestinal cancer [87,88]. They will be the subject of intensive discussion elsewhere in this volume (see, Davison).

3.1.1
Metallocene Complexes – Ti, Mo, Nb, V, Re

Many metallocene(IV) complexes of titanium [89], vanadium [90], niobium [91], tungsten, molybdenum [92] and rhenium have antitumor properties in different tumor systems [93]. They show interesting inhibition effects on the growth of xenografted human carcinomas of the lung, breast and gastrointestinal tract [94]. Vanadocene dichloride [$V(Cp)_2Cl_2$] and, in particular, titanocene dichloride [$Ti(Cp)_2Cl_2$] (Fig. 5a) exhibited the best results.

Ionic metallocene compounds like [$Ti(Cp)_2Cl(NCMe)$][$FeCl_4$] (Fig. 5b), [$Nb(Cp)_2Cl_2$][SbF_6], [$Mo(Cp)_2Cl_2$][SbF_6]$_2$ or [$Re(Cp)_2Cl_2$][AsF_6] are active against Ehrlich ascites tumor cells [95–97]. The titanocene complex salt showed activity against various xenografted human carcinomas. In preclinical experiments mainly hepatotoxicity and gastrointestinal toxicity occurred (the vanadium analogue shows additional nephrotoxicity, due to a different organ distribution). The toxicology and pharmacokinetic parameters of titanocene dichloride have been evaluated in phase I clinical trials [98,99] and it is currently undergoing phase II clinical studies. So far no positive results have been reported.

a   b

Fig. 5. *a* [$Ti(Cp)_2Cl_2$], *b* [$Ti(Cp)_2Cl(NC$-$Me)$][$FeCl_4$]

Titanocene dichloride inhibits DNA synthesis, forms stable adducts and causes DNA damage [100,101].

3.1.2
Bis(β-diketonato)metal Complexes – Ti, Zr, Hf

Some bis(β-diketonato)metal complexes of titanium, zirconium and hafnium, especially those with 1-phenylbutane-1,3-dionato ligands, were found to have significant antitumor activity against a variety of animal tumors like Ehrlich ascites, Stockholm ascitic and sarcoma 180 ascitic tumors, as well as against the solid tumor models MAC 15A colon carcinoma and Walker 256 carcinosarcoma [102–104]. No significant activity was found against the leukemias P388 and L1210. Thus, the compounds are more active in slow growing tumors such as colon tumors than in the fast growing leukemias P388 and L1210.

The most interesting results were obtained in the acetoxymethylmethylnitrosamine-induced colorectal tumor in rats [105]. The titanium(IV) complex diethoxybis(1-phenylbutane-1,3-dionato)titanium(IV), budotitane (Fig. 6), exhibited the best results and was selected for further evaluation.

The zirconium and hafnium complexes are less active than the titanium complexes. Besides variation of the central metal, the bis(β-diketonato) ligand system was systematically modified. As a result the best activity was found with unsubstituted aromatic ring systems [106]. The fact that planar aromatic systems increase antitumor activity suggested intercalation in DNA or RNA strands as the mode of action, but no further proof for this mechanism exists. The quasi-octahedrally configured bis(β-diketonato)metal complexes like budotitane can form three cis- and two trans-isomers, with the cis configuration usually favored. In solution, however, the isomers can convert into each other, with the result that it was not possible to determine which isomer is responsible for the biological activity [107]. The nature of the other ligands does not seem to contribute much to the antitumor activity. Chloride, bromide and fluoride instead of ethoxide (as in the case of budotitane) ligands resulted in no improved activity.

Budotitane became the first non-platinum, transition metal complex to qualify for clinical trials [108–113]. Toxicological evaluation demonstrated hepatoxicity as the main, dose-limiting toxicity. Unfortunately, poor solubility of this ti-

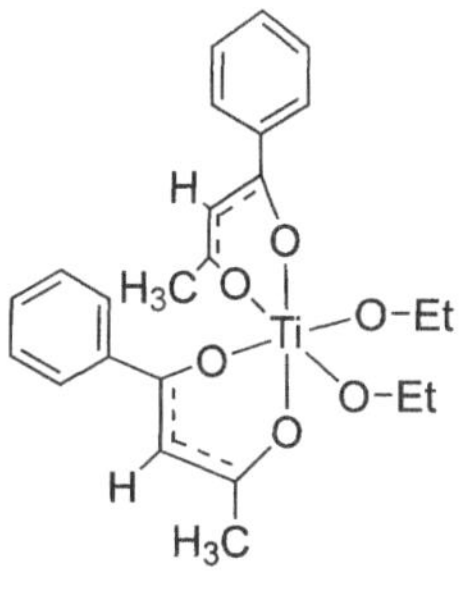

Fig. 6. Budotitane

tanium complex in water made a galenic formulation difficult. Budotitane had to be applied as a co-precipitate, consisting of Cremophor EL, 1,2-propylene glycol and the drug in the ratio of 9:1:1, that gave a micellar solution of the drug in water. Although this solution is stable for several hours and can be used for infusion therapy, analytical documentation of the galenic formulation according to modern standards is not sufficiently possible. Therefore, a subsequent clinical evaluation after phase 1a and 1b clinical studies could not be initiated. Other solutions to this problem and new formulations are on the way.

3.1.3
Vanadium

Vanadate causes inhibition of cell growth in proliferating cultures. Cytotoxicity is dependent on vanadium concentration. The vanadium oxidation state plays a minor role. Inhibition of tumor growth by 80–100% was found in vivo on small MDAY-D2 solid tumors in mice when orthovanadate or vanadyl sulfate was injected subcutaneously at concentrations greater than 5 µM [114]. Addition of hydrogen peroxide potentiated the cytotoxicity of orthovanadate. The mechanism of action of vanadium compounds remains unclear, but intracellular generation of hydroxyl or vanadium radicals may be involved. Among vanadium complexes which show inhibition of cell growth for human nasopharyngeal carcinoma KB cell, the 1,10-phenanthroline (phen) complex $[VO(phen)]^{2+}$ was found to bind to DNA and cleave supercoiled plasmid Col E1 DNA when hydrogen peroxide was added [115]. By means of ESR spin trapping, the authors found that hydroxyl radicals are generated in a pH-dependent manner. VO^{2+} was less effective and no pH optimum was observed for the VO^{2+}/H_2O_2 system.

A vanadium(III) complex with the amino acid L-cysteine exerts a significant anticarcinogenic effect on benzo[a]pyrene-induced leiomyosarcomas in rats. The mean survival time was prolonged and a complete remission of 18% of the tumors was achieved [116].

3.2
Iron Group – Fe, Co, Ni

To date, there have been no promising reports on antiproliferative properties of iron compounds with regard to a clinical application. Iron(III) compounds of the type $[Fe(C_5H_5)_2]X$ are active in vivo against Ehrlich ascites tumor cells [117] but because of hydrolysis and limited activity against xenografted human carcinomas their development was not continued. Interest is mainly focused on the biological activity of the bleomycin-iron complexes [118–122]. As in the case of iron, cobalt and nickel complexes with bleomycin were investigated in relation to their biological activity [123,124].

Inhibition of DNA synthesis of sarcoma 180 tumor cells was reported for cobalt(II) and nickel(II) complexes of 2,2'-diamino-4,4'-bithiazole [125].

The complex salt sodium *trans*-dinitrobis(2,4-pentanedionato)cobalt(III) binds to adenine-rich regions of DNA and causes single-strand scission under irradation of light [126].

Copper(II) and nickel(II) complexes of thiosemicarbazones, thioureas and 2-substituted pyridines have shown to be potent antineoplastic agents in the Ehrlich ascites carcinoma screen [127,128]. In vivo activity in the Ehrlich ascites carcinoma has been reported for nickel(II) thiosemicarbazone complexes which caused L1210 DNA strand scission and inhibited L1210 DNA purine synthesis and several enzyme activities but not L1210 DNA topoisomerase II [129].

2-Hydroxybenzanilinato complexes of cobalt are able to reduce growth of the intramuscularly transplanted Walker 256 carcinosarcoma [130].

Organocobalt(III) compounds with a σ-bonded organyl group show a pH-dependent generation of alkylating agents. Mild acidic conditions generate free alkyl radicals, probably responsible for the significant modification and reduction of rat tumors [131].

Preliminary investigations of the antitumor activity of alkyne-cobalt carbonyl complexes were reported recently. The organometallic compounds are more active against human melanoma and lung carcinoma cell lines than cobalt chloride or dicobalt octacarbonyl [132].

3.3
Platinum Metals – Ru, Rh, Pd (Ir, Re)

Because of their proximity to platinum in the periodic table, complexes of ruthenium, rhodium and palladium were intensively investigated in relation to their antitumor properties. The ruthenium complexes, in particular, are at an advanced stage of preclinical development. The imidazolium salt of the complex *trans*-[RuCl$_4$(Me$_2$SO)im] has recently reached clinical trials (see Sava, this volume).

3.3.1
Ruthenium

The tumor-inhibiting activity of ruthenium complexes such as "ruthenium red", [Ru(NH$_3$)$_3$Cl$_3$], *cis*-[Ru(NH$_3$)$_4$Cl$_2$]Cl or *cis*-[RuCl$_2$(Me$_2$SO)$_4$], has been known for rather a long time [133–135]. The ruthenium(II)-DMSO complex *cis*-[RuCl$_2$(Me$_2$SO)$_4$] showed, together with mild toxicity, better antitumor activity in vivo than cisplatin against Ehrlich ascites carcinoma, Lewis lung carcinoma, B16 melanoma, MCa mammary carcinoma and inhibited the development of pulmonary metastases in the Lewis lung carcinoma [136]. Even superior antitumor and antimetastatic efficacy in the Lewis lung carcinoma model was obtained with the corresponding *trans*-complex *trans*-[RuCl$_2$(Me$_2$SO)$_4$] (Fig. 7a) and the ruthenium(III) complex *mer*-[RuCl$_3$(Me$_2$SO)$_2$(NH$_3$)] [137–139].

The ruthenium(III) complex salt Na *trans*-[RuCl$_4$(Me$_2$SO)im] (im=imidazole) (Fig. 7b) was found to inhibit selectively spontaneous lung metastases in a

a   b

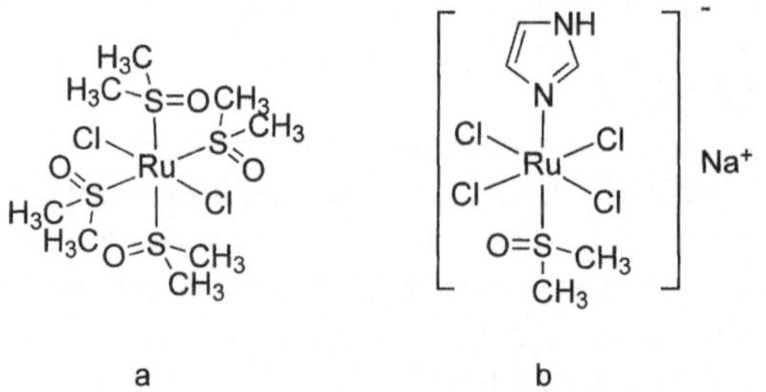

Fig. 7. *a* trans-[RuCl$_2$(Me$_2$SO)$_4$], *b* Na trans-[RuCl$_4$(Me$_2$SO)im]

model of the solid metastasising tumor MCa mammary carcinoma of CBA mice
[140,141]. Compared to cisplatin, it was less active on primary tumor growth.
Comparative investigations on different derivatives of Na trans-[RuCl$_4$(Me$_2$
SO)im] resulted in the conclusion that reduction of metastasis formation is not
correlated to cytotoxicity [142]. Moreover, the activity of Na trans-[RuCl$_4$
(Me$_2$SO)im] seems to be independent of its concentration in tumor cells. It is assumed that this ruthenium(III) drug increases resistance against metastasis formation by potentiation of the extracellular matrix and reduction of blood
stream invasion by tumor cells [143].

A combination treatment of Na trans-[RuCl$_4$(Me$_2$SO)im] and 5-fluoruracil
against the solid metastasizing MCa mammary carcinoma of the CBA mouse
and the lymphocytic leukemia P388 leads to better results than those obtained
with each agent alone [144].

Polypyridyl-ruthenium complexes have been intensively studied as probes of
DNA conformation or DNA cleavage agents [145–147]. The analogous complexes with aqua or chloro ligands as leaving groups bind DNA covalently [148]. The
chloropolypyridyl-ruthenium complex mer-[Ru(terpy)Cl$_3$] (terpy=2,2':6'2"-
terpyridine)] exhibits a significant DNA interstrand cross-linking. mer-
[Ru(terpy)Cl$_3$] is cytotoxic in human cervix carcinoma HeLa and murine L1210
tumor cell lines and shows in vivo antitumor activity in the murine lymphosarcoma LS/BL system [149].

The antitumor effect of the ruthenium(III) complex with 1,2-propylenedi-
aminetetraacetic acid might be due to its ability to stimulate the release of toxic
oxygen metabolites from phagocytic cells infiltrating tumor masses [150,151].

Many ruthenium(III) complexes with heterocyclic and chloride ligands have
been synthesized and screened towards their antitumor potential [152]. As a result, complex salts of the type (HL)$_2$[RuCl$_5$L], and especially HL trans-[RuCl$_4$L$_2$]
with N-bound heterocycles (L), showed the best activity and are much more soluble than the neutral complexes [RuCl$_3$L$_3$].

These compounds exhibited marked activity in vivo against P388 leukemia
with T/C values of 140–200%. For the complex salt HIm trans-[RuCl$_4$(im)$_2$] (im=
imidazole) (Fig. 8a) a T/C value of 200% was obtained and for HInd *trans-*

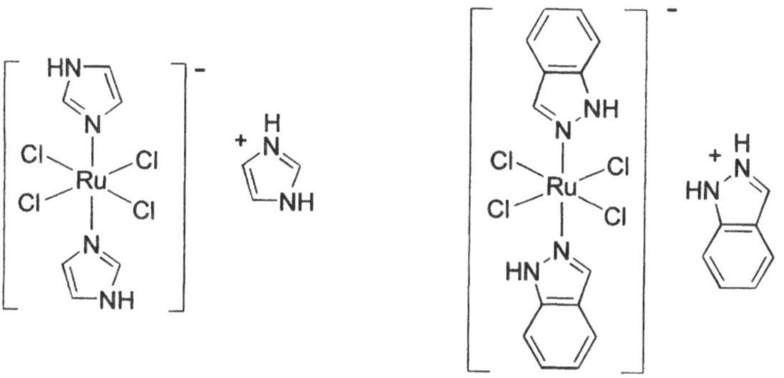

Fig. 8. *a* HIm *trans*-[RuCl$_4$(im)$_2$], *b* HInd *trans*-[RuCl$_4$(ind)$_2$]

[RuCl$_4$(ind)$_2$] (ind=indazole) (Fig. 8b) it was 160% (in comparison: cisplatin: 180%, 5-fluoruracil: 150%). Excellent activity was also detected in Walker 256 carcinosarcoma, Stockholm ascitic tumor, subcutaneously transplanted B16 melanoma, intramuscularly growing sarcoma 180, Ehrlich ascites and MAC 15A colon tumor.

The most interesting results for the imidazole and indazole complexes, with a tumor reduction of 70–90%, were obtained in the acetoxymethylmethylnitro-samine-induced colorectal tumor of the rat [153–157]. This autochthonous colorectal tumor resembles human colon tumors in its histological appearance and behavior against chemotherapeutics. Cisplatin is completely inactive in this model. Furthermore, the two Ru(III) complex salts showed antiproliferative ac-tivity in two human colon cancer cell lines (SW707 and SW948) [158,159]. HInd *trans*-[RuCl$_4$(ind)$_2$], the less toxic complex, and HIm *trans*-[RuCl$_4$(im)$_2$] exhibit antineoplastic effects on proliferation of clonogenic cells from freshly explanted human tumors in a capillary soft agar cloning system in a concentration de-pendent manner [160,161]. Activity was observed against non-small-cell lung, breast and renal cancer.

In the course of formulation of HInd *trans*-[RuCl$_4$(ind)$_2$] for clinical investi-gations, the corresponding sodium salt Na *trans*-[RuCl$_4$(ind)$_2$] was prepared, showing a significantly increased water solubility [162].

The ruthenium(III) complexes have to be seen as prodrugs. The compounds un-dergo hydrolysis reactions under physiological conditions [163–165] and binding towards plasma proteins occurs (see below). Moreover, an "activation by reduction" is assumed for the Ru(III) complexes [166]. They can be activated by reduction to la-bile Ru(II) species. In vivo and in vitro reduction of some Ru(III) compounds by bi-ological reductants has been demonstrated [167,168]. The reduced species are reac-tive compounds that interact with molecular targets like N7 of guanines as preferen-tial binding sites on DNA [169,170]. As solid tumor tissues have a reducing, hypoxic environment, a higher Ru(II)/Ru(III) ratio can be expected than in "normal", more aerated tissues, resulting in a selective cytotoxic effect against tumor cells.

Another important feature of ruthenium complexes, causing selectivity for tumor cells, is their ability to use the transferrin transport cycle. Transferrin is a 80 kDa blood plasma protein that transports iron into the cell. As rapidly growing tumor tissues have a considerable need for iron, tumor cells have an increased amount of transferrin receptors on the cell surface. Radiolabelling experiments with [103]Ru demonstrated an accumulation of a Ru(III)-transferrin adduct in the tumor of mice bearing subcutaneous EMT-6 sarcoma [171]. The ruthenium(III) complexes HInd trans-[RuCl$_4$(ind)$_2$] and HIm trans-[RuCl$_4$(im)$_2$], as well as Na trans-[RuCl$_4$(Me$_2$SO)Im] [172], have been shown to bind to transferrin, specifically at the iron-binding clefts, but also at histidine residues on the surface of the protein [173,174]. The transferrin-bound form of these complexes, prepared by preincubation with the protein, were tested in a human colon cancer cell line. Their antiproliferative activity was retained, or even exceeded, compared to the free complexes [175]. The binding to transferrin enables these ruthenium complexes to accumulate selectively in tumor cells via an indirect tumor targeting.

3.3.2
Rhodium (Iridium, Rhenium)

The tumor-inhibiting properties of dirhodium tetracarboxylates are known. These rhodium compounds showed statistically significant activity in the Ehrlich ascitic tumor, the sarcoma 180 ascitic tumor, and in the intraperitoneally transplanted leukaemias P388 and L1210 [176]. If acetate, propionate, and butyrate are used as carboxylates, an increase in antitumor activity can be observed in the following order: acetate<propionate<butyrate. Hence, antitumor activity increases with the lipophilicity of the compounds.

Dimeric μ-carboxylato rhodium(II) complexes of the type [Rh$_2$(OOCR)$_4$L)$_2$] (Fig. 9a), [Rh$_2$Cl$_2$(OOCR)$_2$(N-N)$_2$] and [Rh$_2$(OOCR)$_2$(N-N)$_2$(H$_2$O)$_2$](OOCR)$_2$ (L=Lewis base; R=CH$_3$, CH$_3$CH(OH), C$_6$H$_5$CH(OH); N-N=1,10-phenantroline

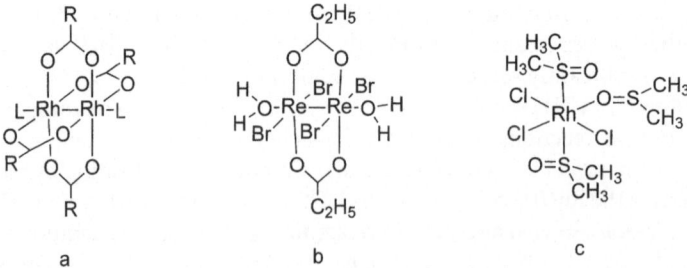

Fig. 9. *a* [Rh$_2$(OOCR)$_4$L)$_2$], *b* [Re$_2${OOC(C$_2$H$_5$)}$_2$(H$_2$O)$_2$Br$_4$], (*c*) *mer,cis*-RhCl$_3$(Me$_2$SO)$_2$ (Me$_2$SO)

and 4,7-dipenyl-1,10-phenantroline) showed activity against the synchronously cultivated green alga chlorella vulgaris [177] and in the human oral carcinoma KB cell line [178]. The cationic complexes, in particular, show higher cytotoxic activity than $[Rh_2(OAc)_4(H_2O)_2]$.

The μ-carboxylatodirhodium(II) complexes easily react with human serum albumin (7 mol per mol protein are bound in the case of $[Rh_2(OOCCH_3)(bpy)_2 (H_2O)_2](OOCCH_3)_2)$, changing the conformation of albumin. The rhodium atoms are coordinated via the imidazole rings of the histidine residues of the protein [179,180].

Structurally related, biologically active μ-carboxylato complexes of ruthenium and rhenium are also known. The rhenium complex $[Re_2\{OOC(C_2H_5)\}_2 (H_2O)_2Br_4]$ (Fig. 9b) shows significant activity in the intraperitoneally transplanted leukaemia P388, in the subcutaneously transplanted sarcoma 180, and in the subcutaneously transplanted melanoma B16. This complex proved to be superior to similar complexes with acetate and butyrate ligands and chloride as halogen ligand [181,182].

Very recently, rhodium(III) analogues of antitumor-active ruthenium(III) compounds (see above), containing imidazole (im) ligands, have been synthesized and tested against the human tumor cell lines A2780 (ovarian carcinoma), A2789/cp8 (cispaltin resistant subline), LoVo (colon carcinoma) and CaLu (lung carcinoma) [183]. The compounds *mer,cis*-RhCl$_3$(Me$_2$SO)$_2$(im) and, in particular, *mer,cis*-RhCl$_3$(Me$_2$SO)$_2$(NH$_3$) showed cytotoxic activity similar to cisplatin, whereas *mer,cis*-RhCl$_3$(Me$_2$SO)(im)$_2$, Na *trans*-[RhCl$_4$(Me$_2$SO)(im)]·2Me$_2$SO and HIm *trans*-[RhCl$_4$(im)$_2$] had no significant effect. In first in vivo tests against the MCa mammary carcinoma, an intramuscular implanted solid tumor of CBA mice which gives spontaneous lung metastases, the rhodium-DMSO compounds were more toxic than the corresponding ruthenium compounds. Only Na *trans*-[RhCl$_4$(Me$_2$SO)(im)]·2Me$_2$SO and the precursor *mer,cis*-RhCl$_3$(Me$_2$SO)$_2$(Me$_2$SO) (Fig. 9c) modified slightly the growth of the primary tumor (25%). Interesting is the fact that *mer,cis*-RhCl$_3$(Me$_2$SO)$_2$(Me$_2$SO) is active against growth of spontaneous metastases comparable to Na *trans*-[RuCl$_4$(Me$_2$SO)(im)] under similar experimental conditions. It reduced 80% of the weight of the metastases, but the decrease is not statistically significant (maybe because of the great variability inside some groups). The reason for the lesser activity of the Rh(III) compounds compared to the Ru(III) compounds might be the relative inertness of the Rh(III) complexes (slower hydrolysis and minor interactions with double stranded DNA were observed) and their lower capacity to be activated by redox reactions.

Activity in different systems in vitro (KB cells) and in vivo (P388 leukemia, Ehrlich ascites carcinoma, sarcoma 180) is known for rhodium(I) dithiocarbamate complexes of the type $[Rh(CO)_2X]$ (X=dithiocarbamate derivative) [184].

3.3.3
Palladium

In general palladium(II) complexes are orders of magnitude more labile than structurally related platinum(II) complexes. For example, [Pd(en)Cl$_2$] (en=1,2-diaminoethane) undergoes spontaneous solvolysis in aqueous solution to produce the mono and diaqua complexes. Although thermodynamic properties like the stability constants are very similar, the rate constant for the hydrolysis of [Pd(en)Cl$_2$] is higher than that of cisplatin by a factor of $2 \cdot 10^5$ [185,186].

Figure 10b shows 1,2-diaminoethanedinitratopalladium(II), [Pd(en)(NO$_3$)$_2$], the basic structure of some tumor-inhibiting palladium complexes. At a dose of 80 mg/kg, the survival time of treated animals (T/C=194%) bearing the sarcoma 180 ascitic tumor was almost doubled [187]. However, the compound was applied intraperitoneally in a single dose on the same day the tumor was transplanted. When applied intraperitoneally the tumor should normally be allowed to grow for at least 24 hours before therapy starts. Other, similarly configured compounds, containing diaminocyclohexane, 1,2-diaminopropane, 1,3-diaminopropane, 2,2'-bipyridine or ammonia as ligands, and malonate, oxalate, glutarate, and cyclobutanedicarboxylate as hydrolisable groups instead of nitrate, were less active under the same conditions and, in part, even inactive.

Some palladium complexes of the type Pd(N-C)LX (N-C is an aromatic or aliphatic amine coordinated as a chelating ligand through the amine and a cyclometallated Pd-C bond; L=amine; X=chloride or acetate, for an example see Fig. 10a) were found to be cytotoxic in a panel of seven human tumor cell lines [188].

Bis(3-hexamethyleneiminyl-2-acetylpyridine-thiosemicarbazonato)palladium(II), among several Pd(II) compounds with 4-N-substituted derivatives of 2-acetylpyridine thiosemicarbazone, is active against leukemia P388. It inhibits incorporation of 3H-thymidine into DNA and induces SCEs and cell division delays [189].

The Pd(II) complex *trans*[Pd(2-dqmp)$_2$Br$_2$] (2-dqmp=diethyl-2-quinolylmethylphosphonate) displayed cytotoxic activity on KB and L1210 cell lines [190].

a   b   **Fig. 10.** *a* Pd(N-C)LX, *b* [Pd(en)(NO$_3$)$_2$]

3.4
Copper Group – Cu, Au

In relation to to their antitumor activity, many interesting compounds of copper and gold have been found. In the case of gold complexes, in particular, redox reactions and stability of oxidation states are important features.

3.4.1
Copper

The Cu(II) complex *trans*-[bis(1,2-hydroxybenzaldoximato)copper(II)] (Fig. 11a) and its derivatives exert antiproliferative activity against Ehrlich ascites tumors in vivo and in vitro [191]. Adverse effects are accumulation of copper in the pancreas and transient blindness.

Promising compounds were obtained with the macrocyclic ligand tetrabenzotetraazacyclohexadecine. Its copper(II) complex (Fig. 11b) reached T/C values of 170% in the melanoma B16 and 140% in the leukaemia P388 [192–194].

Copper(I) and copper(II) complexes with 1,1'-bis(diphenylphosphino)ferrocene and 1,1'-bis(oxodiphenylphosphoranyl)ferrocene were found to exhibit cytotoxic potency against Eagle's KB cell line [195].

Copper(I) complexes with diphenylphospinoethane were tested for cytotoxicity on human ovarian carcinoma cells. Cytotoxicity of the copper(I) compounds was found to be higher compared to adriamycin, but lower compared to cisplatin [196].

Copper complexes of 2-furaldehyde oxime derivatives are cytotoxic agents in murine and human tissue culturered cell lines and inhibit L1210 lymphoid leukemia DNA synthesis followed by RNA synthesis. Purine synthesis regulatory enzyme activities, as well as topoisomerase II activity was markedly reduced [197].

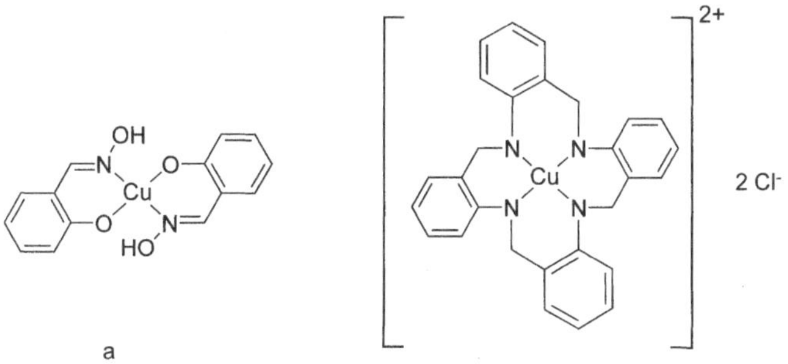

Fig. 11. *a trans*-[Bis(1,2-hydroxybenzaldoximato)copper(II)], *b* [tetrabenzo[*b,f,j,n*](1,5,9,13) tetraazacyclohexadecinecopper(II)] dichloride

A copper benzohydroxamic acid complex was studied against Ehrlich ascites carcinoma in Swiss A mice. The application of 25 mg/kg enhanced life time significantly [198].

Some copper compounds containing organic ligands, that exhibit antiproliferative activity by themselves, are known [199,200]. Besides thiosemicarbazone complexes [201], the complex with bleomycin has been investigated and synthetic analogues of copper(II)-bleomycin synthesized [202,203].

Very recently, the antitumor activity of copper(II) complexes with boron containing ligands has been reported [204]. {N-[(trimethylamineboryl)carbonyl]-L-phenylalanylcarboxylato}-bis-{N-[(trimethylamineboryl)carbonyl]-L-phenylalanine}copper(II) blocked L1210 leukemic cell DNA and RNA synthesis by inhibiting multiple enzyme activities for nucleic acid synthesis. Compared to the free ligand, the copper complex exhibited improved ability to inhibit L1210 partially purified DNA topoisomerase II [205].

Synergistic effects of some Cu(II) complexes have been found in vitro in combination with epirubicin and mitomycin C against HeLa-S3 cervical cancer cell lines [206].

3.4.2
Gold

Gold coordination compounds, especially auranofin, which can be given orally because it is well absorbed by the gastrointestine, are known for their use in the treatment of primary chronic polyarthritis [207] (see Shaw, TBIC, Vol 2). Interest in tumor-inhibiting gold complexes was aroused by the in vitro cytotoxic activity of auranofin (Fig. 12a) and its derivatives. This promising activity in cell cultures could not, however, be reproduced in vivo [208]. Antitumor activity against a wider range of experimental tumors was shown with other gold complexes, particularly those with 1,2-bis(diphenylphosphino)ethane (dppe) as lig-

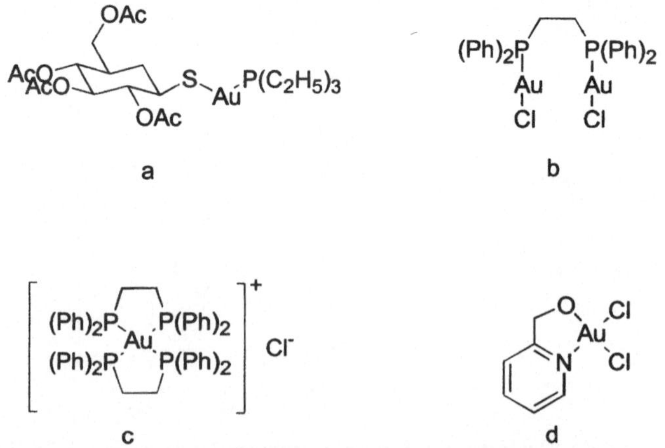

Fig. 12. *a* Auranofin, *b* [ClAu(dppe)AuCl], *c* [Au(dppe)$_2$]Cl, (*d*) AuCl$_2$(pm)

and [209]. The ligand itself also shows antitumor activity, but this is reinforced by the complexation with gold. [ClAu(dppe)AuCl] (Fig. 12b) reaches T/C values of about 200% in the P388 leukaemia. In addition, significant activity was demonstrated in the L1210 leukaemia, the M5076 reticulum cell carcinoma, the B16 melanoma, the mammary adenocarcinoma 16/c (all of these are intraperitoneally transplanted and intraperitoneally treated), and the ADJ/PC6 blastocytoma (subcutaneously transplanted). [Au(dppe)$_2$]Cl (Fig. 12c) [210] is also active, showing comparable activity in the P388 leukaemia, but showed no activity when administered intravenously or subcutaneously. Activity can only be achieved by local application to the tumor. There was also no activity when P388 leukaemia was transplanted intravenously and when the complex was given intraperitoneally or intravenously afterwards. T/C values of about 160% were found in the intraperitoneally transplanted M5076 reticulum cell carcinoma, and T/C values of about 140% were found in the B16 melanoma. Activity was also observed in the subcutaneously transplanted mammary adenocarcinoma 16/c. [Au(dppe)$_2$]Cl causes DNA-protein cross-links and DNA strand breaks, and inhibits protein synthesis more effectively than DNA and RNA synthesis. As a disadvantage, the bis(diphosphine)gold complex has adverse effects on the heart and lungs and uncouples oxidative phosphorylation in mitochondria [211].

Gold(I) complexes with chiral sugar-substructure phosphine ligands of the formula [Au(n-mbpa)L] [n-mbpa=methyl-4,6-O-benzylidene-n-deoxy-n-(diphenylphosphino)-α-D-altropyranoside; L=pyrimidine-2-thione or 3,5-dimethylpyrimidine-2-thione] showed antitumor activity in recent preliminary experiments against P388 leukemia [212].

Activity against Friend leukemia cells is known of the complex (8-thiotheophyllinate)(triphenylphosphine)gold(I). Cytotoxicity is significantly reduced in the presence of albumin or if free thiol groups are accessible, for example, L-cysteine or reduced glutathione [213].

The gold complex potassium[dicyanogold(I)] is active against the Lewis lung carcinoma, reaching T/C values of 200% and producing a high percentage of cured animals if it is applied intraperitoneally from days 1 to 9 at a dose of 2.5 mg/kg. However, comparatively promising activity could not be reproduced in other tumor models and development of this compound was not pursued [214,215].

Cytotoxicity studies of gold(III) complexes of the type AuCl$_3$(Hpm) (Hpm=2-hydroxymethylpyridin) and AuCl$_2$(pm) (Fig. 12d) on a number of tumor cell lines demonstrate good activity in cisplatin-resistant cell lines. A disadvantage of these compounds is their fast hydrolysis and/or reduction under physiological conditions. Their cytotoxic action is probably caused by their metabolites. Interaction with proteins, e.g. albumin or transferrin, leads to hydrolysis products and reduction to gold(I) species. In contrast, the binding to calf thymus DNA or polynucleotides results in adducts with gold remaining in the +3 oxidation state [216].

The gold(III) complexes [Au(N-MeIm)Cl$_3$], [Au(2-MeBo)Cl$_3$] and [Au(2,5-diMeBo)Cl$_3$] (where N-MeIm=N-methylimidazole; 2-MeBO=2-methylbenzoxazole; 2,5-diMeBO=2,5-dimethylbenzoxazole) showed interesting cytotoxicity in vitro in murine leukemia cell lines L1210, human ovarian carcinoma A2780 and their cisplatin-resistant sublines L1210/CDDP and A2780/CDDP, in comparison with cisplatin and carboplatin [217].

Interesting results were reported very recently for square-planar gold(III) complexes, isostructural and isoelectronic with platinum(II) complexes. All these complexes, containing at least two chloride ligands in the cis position, were cytotoxic in different human tumor cell lines. They retained their antitumor activity against the cisplatin-resistant tumor cell lines CCRF-CEM/R leukemia and A2789/R ovarian carcinoma with only minimal cross-resistance with cisplatin [218]. The complexes containing a salicylaldiminate ligand induced tumor cell growth inhibition comparable to or even greater than cisplatin. In spite of the structural relationship with cisplatin and other platinum(II) complexes, the mode of action of these gold(III) complexes might be different. The authors found binding towards calf thymus DNA and alteration of its B-type conformation, but the gold(III) complexes are poorly stable under physiological conditions and have therefore to be seen as prodrugs not reaching the DNA target unchanged. In general, the relatively high redox potential and poor stability of Au(III) complexes are the limiting features of these compounds with regard to further drug development.

4
Perspectives

Many interesting studies into antitumor activity of non-platinum metal antitumor compounds have been published during the last decade. Together with agents developed earlier they have led to new developments in cancer chemotherapy.

However, synthesis of metal complexes with antitumor activity is one concern, developing a pharmaceutical for clinical application is another. Unfortunately, most of the compounds with promising in vitro or even in vivo activity do not overcome preclinical evaluation. This often occurs due to problems in formulation of the drug, including stability of the compound and analytical characterization of resulting products and metabolites. The various reaction opportunities of metal complexes under physiological conditions are in fact the feature that makes the difference to organic pharmaceuticals. Nevertheless, synthesis and evaluation of new antitumor metal compounds remain a challenge to bioinorganic chemists and cooperation with scientists in related fields is worthwhile.

New concepts of tumor targeting might result in drugs with enhanced selectivity for tumor tissue and, as a consequence, higher cure rates and less side effects. This concept of tumor targeting or drug targeting is based on the combination of an antitumor active metal complex with carrier molecules having specific affinity for particular tumor tissues. Some promising examples are known in the field of platinum compounds, for example, *cis*-configurated Pt(II) complexes with a coordinated osteotropic bisphosphonate moiety, which are less toxic than cisplatin and show excellent activity in a transplantable osteosarcoma model [219], or Pt(II) hormone derivatives, which are active in estrogen receptor positive tumors [220]. Coupling of metal complexes with macromolecules like plasma proteins in analogy to known conjugates of organic anticancer drugs is another promising strategy [221]. In the case of some ruthenium or gallium compounds (see above) this drug targeting seems to have already been "naturally" realized.

References

1. Adamson RH, Canellos GP, Sieber SM (1985) Cancer Chemother Rep 59:599
2. Hayes RL, Nelson B, Swartzendruber DC, Carlton JE, Byrd BL (1970) Science 167:289
3. Larson SM, Rasey JS, Allen DR, Nelson NJ, Grunbaum Z, Harp GD, Williams DL (1980) J Natl Cancer Inst 64:41
4. Chitambar CR, Zivkovic Z (1987) Cancer Res 47:3929
5. Chitambar CR, Sax D (1992) Blood 80:505
6. Hegge FN, Mahler DJ, Larson SM (1977) J Nucl Med 18:937
7. Chitambar CR, Werely JP (1997) J Biol Chem 272:12151
8. Chitambar CR, Narasimhan J, Guy J, Sem DS, O'Brien WJ (1991) Cancer Res 51:6199
9. Foster BJ, Clagett-Carr K, Hoth D, Leyland-Jones B (1986) Cancer Treat Rep 70:1311
10. Crawford ED, Saiers JH, Baker LH, Costanzi JH, Bukowski RM (1991) Urology 38:355
11. Seidman AD, Scher HI, Heinemann MH et al. (1991) Cancer 68:2561
12. Warrell RP Jr, Coonley CJ, Straus DJ, Young CW (1983) Cancer 51:1982
13. Warrell RP Jr, Bockman RS, Coonley CJ, Isaacs M, Staszewski H (1984) J Clin Invest 73:1487
14. Jabboury K, Frye D, Holmes FA, Fraschini G, Hortobagyi G (1989) Invest New Drugs 7:225
15. Scher HJ, Curley T, Geller N, Dershaw D, Chan E, Nisselbaum J, Alcock N, Hollander P, Yagoda A (1987) Cancer Treatment Rep 71:887
16. Warrell RP, Israel R, Frisone M, Snyder T, Gaynor JJ et al. (1988) Ann Internal Medicine 108:669
17. Warrell RP, Murphy WK, Schulman P, O'Dwyer PJ, Heller G (1991) J Clin Oncol 19:1467
18. Collery P, Morel M (1990) Oral administration of gallium in conjunction with platinum in lung cancer treatment. In: Collery P, Poirier LA et al. (eds) Metal ions in biology and medicine. John Libbey Eurotext, Paris, p 437
19. Einhorn LH, Roth BS, Ansari R et al. (1994) J Clin Oncol 12:2271
20. Dreicer R, Propert KJ, Roth BJ, Einhorn LH, Loehrer PJ (1997) Cancer 79:110
21. Dreicer R, Lallas TA, Joyce JK, Anderson B, Sorosky JI, Buller RE (1998) Am J Clin Oncol 21:287
22. McCaffrey JA, Hilton S, Mazumdar M, Sadan S, Heineman M, Hirsch J, Kelly WK, Scher HI, Bajorin DF (1997) 15:2449
23. Chitambar CR, Zahir SA, Ritch PS, Anderson T (1997) Am J Clin Oncol 20:173
24. Lundberg JL, Chitambar CR (1990) Cancer Res 50:6466
25. Chitambar CR, Werely JP, Haq RU (1994) Cancer Res 54:3224
26. Hata Y, Sandler A, Loehrer PJ, Sledge GW Jr, Weber G (1994) Oncol Res 6:19
27. Collery P, Millart H, Pechery C, Kratz F, Keppler BK (1992) New gallium complexes for a cisplatin combination therapy. In: Anastassopoulou J, Collery P, Etienne JC, Theophanides T (eds) Metal ions in biology and medicine, Vol 2. John Libbey Eurotext, Paris, p 173
28. Collery P, Domingo JL, Keppler BK (1996) Anticancer Res 16:687

29. Gey DC, Schilling T, Keppler BK, Thiel M, Schmidt-Gayk H, Raue F, Ziegler R (1997) Exp Clin Endocrinol Diabetes 105:6
30. Winter B, Schilling T, Gey DC, Keppler BK, Ziegler R (1998) Exp Clin Endocrinol Diabetes 106:83
31. Rice LM, Slavik M, Schein P (1977) Clinical brochure: spirogermanium (NSC-192965). National Cancer Institute, Bethesda, MD
32. Hill BT, Whatley SA, Bellamy AS, Jenkins LY, Whelan RDH (1982) Cancer Res 42:2852
33. Whelan RD, Hill BT (1982) Br J Cancer 45:639
34. Goodwin JW, Kopecky K, Slavik M, Tranum BL, Balcerzak SP, Fletcher WS, Constanzi JJ (1987) Cancer Treatment Rep 71:985
35. Slavik M, Elias L, Mrema J, Saiers JH (1982) Drugs Exptl Clin Res 8:379
36. Slavik M, Blanc O, Davis J (1983) Invest New Drugs 1:225
37. Ettinger DS, Finkelstein DM, Abeloff MD, Chang YC, Smith TJ, Oken MM, Ruckdeschel JC (1990) Invest New Drugs 8:183
38. McMaster ML, Greco FA, Johnson DH, Hainsworth JD (1990) Invest New Drugs 8:87
39. Sato H, Iwaguchi A (1979) Jap J Cancer Chemother 6:79
40. Kumano N, Ishikawa T, Koinamura S, Kikumoto T, Suzuki S, Nakai Y, Konno K (1985) Tohuku J Exp Med 146:97
41. Suzuki F, Pollard RB (1984) J Interferon Res 4:223
42. Ikemoto K, Kobayashi M, Fukumoto T, Morimatsu M, Pollard RB, Suzuki F (1996) Experientia 52:159
43. Nakada Y, Kosaka T, Kuwabara M, Tanaka S, Sato K, Koide F (1993) J Vet Med Sci 55:795
44. Hopkins SJ (1980) Drugs of the Future 5:545
45. Kumano N, Nakai Y, Ishikawa T, Koinamura S, Suzuki S, Kikumoto T, and Konno (1980) In: Nelson JD, Grassi CD (eds) Current chemotherapy and infectious disease, Proc 11th Int Congr Chemother, American Society for Microbiology, Washington, p 1525
46. Miyao K, Onishi T, Asai K, Tomizawa S, Suzuki F (1980) In: Nelson JD, Grassi CD (eds) Current chemotherapy and infectious disease, Proc 11th Int Congr Chemother, American Society for Microbiology, Washington, p 1527
47. Sugiya Y, Sugita T Sakamaki S, Abo Y, Satoh H (1986) Oyo Yakuri 23:93
48. Suzuki F, Brutkiewicz RR, Pollard RB (1985) Br J Cancer 757
49. Suzuki F, Brutkiewicz RR, and Pollard RB (1985) Anticancer Res 5:479
50. Tsutsui M, Kakimoto N, Axtell DD, Oikawa H, Asaki K (1976) J Am Chem Soc 98:8287
51. Ward SG, Taylor RC (1988) Antitumor activity of the main-group elements: aluminum, gallium, indium, thallium, germanium, lead, antimony and bismuth. In: Gielen M (ed) Metal-based antitumor drugs. Freund Publishing House Ltd, London, p 1
52. Sato I, Yuan BD, Nishimura T, Tanaka N (1985) J Biol Resp Modif 4:159
53. Köpf-Maier P, Janiak C, Schumann H (1988) J Cancer Res Clin Oncol 114:502
54. Crowe AJ (1993) Tin analogues of cisplatin. In: Keppler BK (ed) Metal complexes in cancer chemotherapy. VCH, Weinheim, p 369
55. Crowe AJ, Smith PJ, Atassi G (1980) Chem Biol Interact 32:171
56. Crowe AJ, Smith PJ, Atassi G (1984) Inorg Chim Acta 93:179
57. Crowe AJ, Smith PJ (1980) Chem Ind 200
58. Gielen M, Joosen E, Mancilla T, Jurkschat K, Willem R, Roobol C, Bernheim J, Atassi G, Huber F, Hoffmann E, Preut H, Mahieu B (1987) Main Group Met Chem 10:147
59. Willem R, Biesemans M, Boualam M, Delmotte A, El Khloufi A, Gielen M (1993) Appl Organomet Chem 7:311
60. Gielen M (1994) Metal-Based Drugs 1:213
61. Gielen M, Biesemans M, El Khloufi A, Meunier-Piret J, Kayser F, Willem R (1993) J Fluorine Chem 64:279
62. Gielen M, El Khloufi A, Biesemans M, Bouhdid A, de Vos D, Mahieu B (1994) Metal-Based Drugs 1:305

63. Crowe AJ (1988) in: Gielen MF (ed) Metal-based anti-tumor drugs. Freund Publishing House, London, p 103
64. Crowe AJ, Smith PJ, Atassi G (1980) Chem Biol Interact 32:171
65. Ruisi G, Silvestri A, LoGiudice MT, Barbieri R, Lamartina L, Atassi G, Huber F, Gräatz K (1985) J Inorg Biochem 25:229
66. Tiekink ERT, Gielen M, Bouhdid A, Willem R, Bregadze VI, Ermanson LV, Glazum SA (1997) Metal-Based Drugs 4:75
67. Mervis J (1996) Science 273:578
68. Chen GQ, Zhu J, Zhang TD,et al. (1996) Blood 88:1052
69. Shao W, Fanelli M, Ferrara FF et al. (1998) J Natl Cancer Inst 90:124
70. Chen GQ, Shi XG, Zhang TD et al. (1997) Blood 89:3345
71. Shen ZX, Chen GQ, Zhang TD et al. (1997) Blood 89:3354
72. Gianni M, Koken MHM, Chelbi-Alix MK, Benoit G, Lanotte M, Chen Z, De The H (1998) Blood 91:4300
73. König A, Wrazel L, Warrel RP jr, Rivi R, Pandolfi PP, Jakubowski A, Gabrilove JL (1997) Blood 90:562
74. Rivi R, Calleja E, König A, Lai L, Gambacorti-Passerine C, Scheinberg D, Gabrilove JL, Warrell RP Jr, Pandolfi PP (1996) Blood 88:68a
75. Soignet S, Rivi R, Tong WP, Gabrilove J, Scheinberg DA, Pandolfi PP, Warrell RP Jr (1997) Proc Am Soc Clin Oncol 16:4a
76. Köpf-Maier P, Klapötke T (1988) Inorg Chim Acta 152:49
77. Dittes U, Diemer R, Keppler BK (1995) J Cancer Res Clin Oncol 121:A50
78. Pfestorf S, Depenbrock H, Keppler BK, Hanauske AR (1996) In: Havemann K, Wolf M (eds) 22nd Deutscher Krebskongreß Abstract Volume. OmniMed, Hamburg, p 164
79. Kondo Y, Satoh M, Imura N, Akimoto M (1991) Cancer Chemother Pharmacol 29:19
80. Kondo Y, Satoh M, Imura N, Akimoto M (1992) Anticancer Res 12:2303
81. Silvestru C, Socaciu C, Bara A, Haiduc I (1990) Anticancer Res 10:803
82. Bara A, Socaciu C, Silvestru C, Haiduc I (1991) Anticancer Res 11:803
83. Socaciu C, Bara A, Silvestru C, Haiduc I (1991) In Vivo 5:425
84. Keppler BK, Silvestru C, Haiduc I (1994) Metal-Based Drugs 1:73
85. Curley T, Scher H, Thaler H, Yeh S, O'Dell M, Kher U, Friedlander-Klar H, Larson S; Foley K, Portenoy R (1992) Proc Ann Meet Am Soc Clin Oncol 11:A672
86. de Klerk JM, van het Schip AD, Zonnenberg BA, van Dijk A, Quirijnen JM, Blijham GH, van Rijk PP (1996) J Nucl Med 37:244
87. Juweid M, Sharkey RM, Swayne LC, Griffiths GL, Dunn R, Goldenberg DM (1998) J Nucl Med 39:34
88. Jakobs AJ, Fer M, Su FM, Breitz H, Thompson J, Goodgold H, Cain J, Heaps J, Weiden P (1993) Obstet Gynecol 82:586
89. Köpf H, Köpf-Maier P (1979) Angew Chem 91:509
90. Köpf-Maier P, Köpf H (1979) Z Naturforsch 34b:805
91. Köpf-Maier P, Leitner M, Köpf H (1980) J Inorg Nucl Chem 42:1789
92. Köpf-Maier P, Leitner M, Voigtländer R, Köpf H (1979) Z Naturforsch 34c:1174
93. Köpf-Maier P, Köpf H (1988) Struct Bond 70:103
94. Köpf-Maier P (1989) Cancer Chemother Pharmacol 23:225
95. Köpf-Maier P, Klapötke T (1992) Cancer Chemother Pharmacol 29:361
96. Köpf-Maier P, Klapötke T (1992) J Cancer Res Clin Oncol 118:216
97. Köpf-Maier P, Neuse E, Klapötke T, Köpf H (1989) Cancer Chemother Pharmacol 24:23
98. Christodoulou C, Ferry D, Fyfe D, Young A, Doran J, Sass G, Eliopoulos A, Sheehan T, Kerr DJ (1997) Proc Ann Meet Am Assoc Cancer Res 38:A1495
99. Berdel WE, Schmoll HJ, Scheulen ME, Korfel A, Grundel O, Harstrick A, Knoche M, Fels LM, Bach F, Baumgart J, Sass G (1995) Proc Ann Meet Am Soc Clin Oncol 14:A1512
100. McLaughlin ML, Cronan JM, Schaller TR, Snelling RD (1990) J Am Chem Soc 112:8949

101. Christodoulou C, Eliopoulos AG, Young LS, Hodgkins L, Ferry DR, Kerr DJ (1998) Br J Cancer 77:2088
102. Keppler BK, Berger MR, Heim ME (1990) Cancer Treat Rev 17:261
103. Keppler BK, Diez A, Seifried V (1985) Drug Res 35:1832
104. Keppler BK, Friesen C, Moritz HG, Vongerichten H, Vogel E (1991) Struct Bond 78:97
105. Bischoff H, Berger MR, Keppler BK, Schmähl D (1987) J Cancer Res Clin Oncol 113:446
106. Keppler BK, Heim ME (1988) Drugs of the Future 13:637
107. Keppler BK, Friesen C, Vongerichten H, Vogel E (1993) Budotitane, a new tumor-inhibiting titanium compound. In: Keppler BK (ed) Metal complexes in cancer chemotherapy. VCH, Weinheim, p 187
108. Heim ME, Keppler BK (1987) J Cancer Res Clin Oncol 113:46
109. Keppler BK, Bischoff H, Berger MR, Heim ME, Reznik G, Schmähl D (1988) Preclinical development and first clinical studies of budotitane. In: Nicolini M (ed) Proceedings of the 5th international symposium on platinum and metal coordination compounds in cancer chemotherapy. Martinus Nijhoff Publishing, Boston, p 684
110. Keppler BK, Heim ME, Flechtner H, Wingen F, Pool BL (1989) Drug Res 39:706
111. Heim ME, Bischoff H, Keppler BK (1990) Clinical studies with budotitane. In: Collery P, Poirier LA, Manfait M, Etienne JC (eds) Metal ions in biology and medicine, John Libbey Eurotext, Paris, p 508
112. Müller U, Schilling T, Keppler BK, Heim ME, Burk K, Rastetter J, Hanauske AR (1994) J Cancer Res Clin Oncol 120:R13
113. Schilling T, Keppler BK, Heim ME, Niebch G, Dietzfelbinger H, Rastetter J, Hanauske AR (1996) Invest New Drugs 13:327
114. Cruz TF, Morgan A, Min W (1995) Proc Ann Meet Am Assoc Cancer Res 36:A2357
115. Sakurai H, Tamura H, Okatani K (1995) Biochem Biophys Res Commun 206:133
116. Evangelou A, Karkabounas S, Kalpouzos G, Malamas M, Liasko R, Stefanou D, Vlahos AT, Kabanos TA (1997) Cancer Lett 119:221
117. Köpf-Maier P, Köpf H, Neuse EW (1984) J Cancer Res Clin Oncol 108:336
118. Lehmann TE, Ming lj, Rosen ME, Que L Jr (1997) Biochemistry 36:2807
119. Sugiura Y, Guan LL (1993) Supramol Chem 1:313
120. Duff RJ, de Vroom E, Geluk A, Hecht SM, van der Marel GA, van Boom JH (1993) J Am Chem Soc 115:3350
121. Carter BJ, de Vroom E, Long EC, van der Marel GA, van Boom JH, Hecht SM (1990) Proc Natl Acad Sci USA 87:9373
122. Kikuchi H, Tetsuka T (1992) J Antibiot 45:548
123. Byrnes RW, Templin J, Sem D, Lyman S, Petering DH (1990) Cancer Res 50:5275
124. Guan LL, Kuwahara J, Sugira Y (1993) Biochemistry 32 6141
125. Thian VN, Yang P, Wang HF, Feng YL, Liu SX, Huang JL Polyhedron (1996) 15:2771
126. Tamura H, Fuijita H (1997) Chem Lett 711
127. Hall IH, Rajendran KG, West DX, Liberta AE (1993) Anti-Cancer Drugs 4:231
128. West DX, Liberta AE, Rajendran KG, Hall IH (1993) Anti-Cancer Drugs 4:241
129. Hall IH, Miller MC III, West DX (1997) Metal-Based Drugs 4:89
130. Hodnett EM, Moore CH, French FA (1971) J Med Chem 14:1121
131. Osinsky SP, Levitin IY, Bubnovskaya LN, Kormuta NA, Ganusevich II, Tsikalova MV, Vol'pin ME (1997) Med Biol Environ 25:75
132. Jung M, Kerr DE, Senter PD (1997) Arch Pharm 330:173
133. Anghileri LJ (1975) Z Krebsforschung 83:213
134. Clarke MJ (1980) The potential of ruthenium in anticancer pharmaceuticals. In: Martell EA (ed) Inorganic chemistry in biology and medicine. American Chemical Society, Washington, p 157
135. Giraldi T, Sava G, Bertoli G, Mestroni G, Zassinovich G (1977) Cancer Res 37:2662
136. Sava G, Giraldi T, Mestroni G, Zassinovich G (1983) Chem Biol Interact 45:1
137. Sava G, Pacor S, Zorzet S, Alessio E, Mestroni G (1989) Pharmacol Res 21:617

138. Mestroni G, Alessio E, Calligaris M, Attia WM, Quadrifoglio F, Cauci S, Sava G, Zorzet S, Pacor S, Monti-Bragadin C, Tamaro M, Dolzani L (1989) Progr Clin Biochem 10:73

139. Pacor S, Sava G, Ceschia V, Bregnant F, Mestroni G, Alessio E (1991) Chem Biol Interact 78:223

140. Sava G, Pacor S, Coluccia M, Mariggio M, Cocchietto M, Alessio E, Mestroni G (1994) Drug Invest 8:150

141. Sava G, Pacor S, Mestroni G, Alessio E (1992) Clin Exp Metastasis 10:273

142. Sava G, Pacor S, Bergamo A, Cocchietto M, Mestroni G, Alessio E (1995) Chem Biol Interact 95:109

143. Bergamo A, Cocchietto M, Capozzi I, Mestroni G, Alessio E, Sava G (1996) Anticancer Drugs 7:697

144. Coluccia M, Sava G, Salerno G, Bergamo A, Pacor S, Mestroni G, Alessio E (1995) Metal-Based Drugs 2:195

145. Barton JK (1986) Science 233:727

146. Grover N, Welch TW, Fairley TA, Cory M, Thorp HH (1994) Inorg Chem 33:3544

147. Gupta M, Grover N, Neyhart GA, Singh P, Thorp HH (1993) Inorg Chem 32:310

148. Grover N, Gupta N, Thorp HH (1992) J Am Chem Soc 114:3390

149. Novakova O, Kasparkova J, Vrana O, van Vliet PM, Reedijk J, Brabec V (1995) Biochemistry 34:12369

150. Vilaplana R, Romero MA, Quiros M, Salas JM, Gonzalez-Vilchez F (1995) Metal-Based Drugs 2:211

151. Carballo M, Vilaplana R, Marquez G, Conde M, Bedoya FJ, Gonzalez-Vilchez F, Sobrino F (1997) Biochem J 328:559

152. Keppler BK, Lipponer KG, Stenzel B, Kratz F (1993) New tumor-inhibiting ruthenium complexes. In: Keppler BK (ed) Metal complexes in cancer chemotherapy. VCH, Weinheim, p 187

153. Seelig M, Berger MR, Keppler BK, Schmähl D (1990) Efficacy of two ruthenium complexes against chemically induced autochthonous colorectal carcinoma in rats. In: Collery P, Poirier LA, Manfait M, Etienne JC (eds) Metal ions in biology and medicine. John Libbey Eurotext, Paris, p 476

154. Berger MR, Seelig MH, Galeano A (1993) Metal complexes with specific activity against colorectal tumors: evaluation of a tumor model close to the clinical situation. In: Keppler BK (ed) Metal complexes in cancer chemotherapy. VCH, Weinheim, p 187

155. Keppler BK, Berger MR, Heim ME (1990) Cancer Treatment Rev 17:261

156. Berger MR, Garzon FT, Keppler BK, Schmähl D (1989) Anticancer Res 9:761

157. Seelig MH, Berger MR, Keppler BK (1992) J Cancer Res Clin Oncol 118:195

158. Galeano A, Berger MR, Keppler BK (1992) Arzneimittelforschung/Drug Research 42:821

159. Kreuser ED, Keppler BK, Berdel WE, Piest A, Thiel E (1992) Semin Oncol 19:73

160. Depenbrock H, Schmelcher S, Peter R, Keppler BK, Weirich G, Block T, Rastetter J, Hanauske AR (1997) Eur J Cancer 33:2404

161. Depenbrock H, Aumüller K, Peter R, Keppler BK, Schweighart S, Block T, Rastetter J, Hanauske AR (1997) J Cancer Res Clin Oncol 121:A51

162. Pieper T, Hartmann M, Sommer M, Keppler BK (1997) J Cancer Res Clin Oncol 123:S35

163. Chatlas J, van Eldik R, Keppler BK (1995) Inorg Cim Acta 233:59

164. Ni Dhubhghaill OM, Hagen WR, Keppler BK, Lipponer KG, Sadler P (1994) J Chem Soc Dalton Trans 3305

165. Anderson C, Beauchamp AL (1995) Can J Chem 73:471

166. Clarke MJ (1989) Prog Clin Biochem Med 10:25

167. Clarke MJ, Bitler S, Rennert D, Buchbinder M, Kelman AD (1980) J Inorg Biochem 12:79

168. Hartmann M, Lipponer KG, Keppler BK (1998) 267:137

169. Clarke MJ, Jansen B, Marx KA, Kruger R (1986) Inorg Chim Acta 124:13

170. Frasca D, Ciampa J, Emerson J, Umans RS, Clarke MJ (1996) Metal-Based Drugs 3:197

171. Srivastava SC, Mausner LF, Clarke MJ (1989) In: Progress in clinical biochemistry and medicine. Springer, Berlin Heidelberg New York, 10:111
172. Messori L, Kratz F, Alessio E (1996) Metal-Based Drugs 3:1
173. Kratz F, Hartmann M, Keppler BK, Messori L (1994) J Biol Chem 269:2581
174. Smith CA, Sutherland-Smith AJ, Keppler BK, Kratz F, Baker EN (1996) J Bioinorg Chem 1:424
175. Kratz F, Keppler BK, Hartmann M, Messori L, Berger MR (1996) Metal-Based Drugs 3:15
176. Erck A, Rainen L, Whilleyman J, Chang JM, Kimball AP, Bear J (1974) Proc Soc Exp Biol Med 145:1278
177. Pruchnik FP, Kluzewska G, Wilczock A, Mazurek U (1997) J Inorg Biochem 65:25
178. Pruchnik F, Dus D (1996) J Inorg Chem 61:55
179. Trynda L, Pruchnik F (1995) J Inorg Biochem 58:69
180. Trynda L, Pruchnik F (1997) J Inorg Biochem 66:187
181. Dimitrov NV, Eastland GW (1977) In: International Congress on Chemotherapy Proceedings of the 10th Current Chemotherapy, p 1319
182. Eastland GW, Yang G, Thompson T (1983) Meth Find Exptl Clin Pharmacol 5:435
183. Mestroni G, Alessio E, Sessanta o Santi A, Geremia S, Bergamo A, Sava G, Boccarelli A, Schettino A, Coluccia M (1998) Inorg Chim Acta 273:62
184. Craciunescu DG, Scarcia V, Furlani A, Papaioannou A, Parrondo IE, Alonso MP (1991) In Vivo 5:329
185. Hohmann H, van Eldik R (1990) Inorg Chim Acta 174:87
186. Hohmann H, Suvachittanont S, van Eldik R (1990) Inorg Chim Acta 177:51
187. Gill DS (1984) Develop Oncol 17:267
188. Higgins JD, Neely L, Fricker S (1993) J Inorg Biochem 49:149
189. Papageorgiou A, Iakovidou Z, Mourelatos D, Miglou E, Boutis L, Kotsis A, Kovala-Demertzi D, Domapoulou A, West DX, Demertzis MA (1997) Anticancer Res 17:247
190. Tusek-Bozic L, Matijasic I, Bocelli G, Calestani G, Furlani A, Scarcia V, Papaioannou A (1991) J Chem Soc Dalton Trans 195
191. Elo H, Lumme P (1987) Inorg Chim Acta 136:149
192. Elo H, Lumme P (1985) Cancer Treatment Rep 69:1021
193. Lumme P, Elo H, Jänne J (1984) Inorg Chim Acta 92:241
194. Lumme P, Elo H (1985) Inorg Chim Acta 107:L15
195. Scarcia V, Furlani A, Philloni G, Corain B (1997) Inorg Chim Acta 254:199
196. Adwanker MK, Wycliff C, Samuelson A (1997) Indian J Exp Biol 35:810
197. Hall IH, Taylor K, Miller MC III, Dothan X, Khan MA, Bouet FM (1997) Anticancer Res 17:2411
198. Khanam JA, Bag SP, Sur B, Sur P (1997) Indian J Pharmacol 29:157
199. Petering DH (1980) Carcinostatic copper complexes. In: Sigel H (ed) Metal ions in biological systems. Marcel Dekker, New York, p 197
200. Umezawa H, Takita T (1980) Struct Bonding 40:73
201. West DX, Ingram JJ, Kozub NM, Bain GA, Liberta AE (1996) Transition Met Chem 21:213
202. Oikawa T, Hirotani K, Ogasawara H, Katayama T, Ashinofuse H, Shimamura M, Iwaguchi T, Takamura O (1990) Chem Pharm Bull 38:1790
203. Brown SJ, Hudson SE, Stephan DW, Mascharak PK (1989) Inorg Chem 28:468
204. Miller MC III, Sood A, Spielvogel BF, Hall IH (1998) Appl Organomet Chem 12:87
205. Miller MC III, Sood A, Spielvogel BF, Hall IH (1998) Metal-Based Drugs 5:1
206. Geromichalos GD, Katsoulos GA, Hadjikostas CC, Kortsaris AH, Kyriakidis DA (1996) Anti-Cancer Drugs 7:469
207. Sutton BM, Franz RG (eds) (1983) Bioinorganic chemistry of gold coordination compounds. Smith Kline & French Laboratories, Philadelphia
208. Ni Dhubhghaill OM, Sadler PJ (1993) Gold complexes in cancer chemotherapy. In: Keppler BK (ed) Metal complexes in cancer chemotherapy. VCH, Weinheim, p 222

209. Lewis AJ, Walz DT (1982) Immunopharmacology of gold. In: Ellis GP, West GB (eds) Progr Medicinal Chem 19. Elsevier Biomedical Press, Lausanne
210. Berners-Price SJ, Girard GR, Hill DT, Sutton BM, Jarrett PS, Faucette LF, Johnson RK, Mirabelli CK, Christopher K, Sadler PJ (1990) J Med Chem 33:1386
211. Sadler PJ (1991) Adv Inorg Chem 36:1
212. Shi JC, Chen LJ, Huang XY, Wu DX, Kang BS (1997) J Organomet Chem 535:17
213. Garcia-Orad A, Arizti P, Sommer F, Silvestro L, Massiot P, Chevallier P, Gutierrez-Zorrilla JM, Colacio E, Martinez de Pancorbo M, Tapiero H (1993) Biomed Pharmacother 47:363
214. Berners-Price SJ, Mirabelli CK, Johnson RK, Mattern MR, McCabe FL, Faucette LF, Sung C-m, Mong S-M, Sadler PJ, Crooke ST (1986) Cancer Res 46:5486
215. Berners-Price SJ, Sadler PJ (1988) Structure and Bonding 70:28
216. Calami P, Carotti S, Guerri A, Messori L, Mini E, Orioli P, Sperio GP (1997) J Inorg Biochem 66:103
217. Cossu F, Matovic Z, Radanovic D, Ponticelli G (1994) Farmaco 49:301
218. Calamai P, Carotti S, Guerri A, Mazzei T, Messori L, Mini E, Orioli P, Speroni GP (1998) Anti-Cancer Drug Des 13:67
219. Klenner T, Valenzuela Paz P, Amelung F, Münch H, Zahn H, Keppler BK, Blum H (1993) Platinum phosphonato complexes with particular activity against bone malignancies. In: Keppler BK (ed) Metal complexes in cancer chemotherapy. VCH, Weinheim, p 85
220. von Angerer E (1993) Platinum complexes with specific activity against hormone-dependent tumors. In: Keppler BK (ed) Metal complexes in cancer chemotherapy. VCH, Weinheim, p 73
221. Kratz F, Fichtner I, Beyer U et al. (1997) Eur J Cancer 33:S175